# AGRICULTURAL CHEMISTRY

Edited by **Margarita Stoytcheva**
and **Roumen Zlatev**

**INTECHOPEN.COM**

**Agricultural Chemistry**
http://dx.doi.org/10.5772/55163
Edited by Margarita Stoytcheva and Roumen Zlatev

**CBS, Edition 2016**

**Contributors**

Roland Solecki, Abdelkarim Abdellaue, Teresa Borges, Kaija Kallio-Mannila, Herbert Köpp, Thierry Mercier, Vera Ritz, Gabriele Schöning and José Tarazona, Laura Echarte, Lujan Nagore, Javier Di Matteo, Matías Cambareri, Mariana Robles and Aída Della Maggiora, Bassam T. Yasseen and Roda F. Al-Thani, Kelly Cristina Tonello and José Teixeira Filho, Ladislav Bláha and Kateřina Pazderů, Yin Gong, Liqun Rao and Diqiu Yu, C. Asha Poorna, M.S. Resmi and E.V. Soniya, Nor Haslinda Hanim Bt Khalil and Tan Guan Huat, Gerald M. Ghidiu, Erin M. Hitchner and Melvin R. Henninger, Ogbonda G. Echem and L. G. Kabari

**Published by InTech**

Janeza Trdine 9, 51000 Rijeka, Croatia


**Publishing Process Manager** Marija Radja

**Technical Editor** InTech DTP team

**Cover** InTech Design team

Additional hard copies can be obtained from orders@intechopen.com

Agricultural Chemistry, Edited by Margarita Stoytcheva and Roumen Zlatev
p. cm.
ISBN 978-953-51-1026-2

INTECH

open science | open minds

# Contents

# Preface

The present book is a collection of 10 original research articles and reports, associated with selected topics in agricultural chemistry. The discussed issues are organized in four sections: Classification and labeling of active substances in plant protection products, Environmental and stress plant physiology and behavior, Antimicrobial and antioxidant potential of plant extracts, and Pollutants analysis and effects.

The first book section (Chapter 1) reports the results of the Workshop held in Berlin on 12-13 April 2011 on harmonized classification and labeling of active substances in plant protection products under the Regulation (EC) No 1272/2008, and the discussion of scientific and practical issues in the interpretation of carcinogenicity, mutagenicity, and reproductive toxicity studies according to Regulations (EC) No 1107/2009 and 1272/2008.

The second book section (Chapters 2-6) addresses a large spectrum of subjects including: grain yield of Argentinean maize hybrids, resource use efficiency and hybrids tolerance to various stresses; ecophysiological studies and problems of restoration and phytoremediation of native plants adapted to various extreme environmental conditions in Qatar; ecophysiological behavior of Eucalyptus grandis x Eucalyptus urophylla at the leaf level in association with environmental variables to develop models predicting the ecophysiological responses; and abiotic stress factors, effects, and tolerance in plants.

The third book section (Chapter 7) focuses on the antimicrobial and antioxidant potential of plant extracts, namely this of the leaves of the mangrove plant *Excoecaria agallocha* Linn. The strong oxidative DNA damage preventive activity and radical scavenging activity of *E. agallocha* Linn is associated with its rich content of flavonoids.

The last book section (Chapters 8-10) is dedicated to the pollutants analysis and effects. Chapter 8 demonstrates that high performance liquid chromatography coupled with mass spectrometry and atmospheric interfaces could be successfully applied for the quantification of triazole fungicides in fruits and vegetables. Chapter 9 describes the effect of simulated rainfall on the control of Colorado Potato Beetle and Potato Leafhopper with At-Plant applications of three neonicotinoid insecticides: imidacloprid, thiamethoxam, and dinotefuran on potatoes in laboratory and field trials. Finally, Chapter 10 reports results on heavy metal content in bitter leaf herbs.

The information provided in this book should be of interest for academic researchers and for agriculturalists.

All the contributing authors are gratefully acknowledged for their time and efforts in preparing the different chapters, and for their interest in the present project.

**Margarita Stoytcheva and Roumen Zlatev**
Mexicali, Baja California
Mexico

Section 1

# Classification and Labeling of Active Substances in Plant Protection Products

# Chapter 1

# Report of the Workshop on Harmonized Classification and Labelling (CLH) of Active Substances in Plant Protection Products Held in Berlin on 12 and 13 April 2011

Roland Solecki, Abdelkarim Abdellaue, Teresa Borges, Kaija Kallio-Mannila, Herbert Köpp, Thierry Mercier, Vera Ritz, Gabriele Schöning and José Tarazona

Additional information is available at the end of the chapter

http://dx.doi.org/10.5772/55603

## 1. Introduction

For approval of active substances Regulation (EC) No 1107/2009 (here referred to as PPP Regulation) provides in Annex II "Procedure and criteria for the approval of active substances, safeners and synergists pursuant to Chapter II" that, amongst other things, active substances, safeners and synergists (here referred to as active substances) cannot be approved if they are classified or have to be classified for carcinogenicity, mutagenicity or reproductive toxicity (CMR), category 1A or 1B hazard classes in accordance with the Classification Labelling and Packaging (CLP) Regulation, unless exposure is negligible (for C and R, 1A or 1B). The PPP Regulation specifies in the approval procedures that the applicant shall submit a dossier to the Rapporteur Member State (RMS), who shall assess the dossier and present the results of that assessment in the draft assessment report (DAR). The RMS shall submit its DAR to the Commission and the European Food Safety Authority (EFSA). EFSA is required to make the DAR available within 30 days to the other Member States for a 60-day commenting period. In parallel, the DAR is also made publicly available by EFSA. EFSA have to adopt a conclusion within 120-150 days of the end of the commenting period on whether the active substance can be expected to meet the approval criteria and send this to the Commission and Member States. The Commission then has to present a review report and a draft regulation (proposed decision) to the Standing Committee on the Food Chain and Animal Health within 6 months of receipt of the conclusion.

For classification of active substances Regulation (EC) No 1272/2008 (here referred to as CLP Regulation), requires that proposals for Harmonized Classification and Labelling (C&L) of active substances in PPP should be submitted to the European Chemicals Agency (ECHA). The proposals follow an agreed procedure with an initial accordance check which is followed by a public consultation process and subsequent consideration of the proposal by the Committee for Risk Assessment (RAC). The legislation requires the RAC to adopt an opinion on the proposal.

The comments received during public consultation may have an impact on the subsequent steps of the process. In a dialogue between the dossier submitter, RAC rapporteurs and ECHA secretariat the best way to proceed will be decided in cases where substantial comments and/or new information are received during the public consultation. In certain cases this may lead to the withdrawal of the dossier and the submission of a revised version by the Member State or to another public consultation on a re-submitted dossier based on the RAC opinion. In other cases the RAC may indicate that the submitted information is insufficient and that it does not allow an opinion to be issued on the classification and labelling.[1]

The legislation requires the RAC to adopt an opinion on the proposal within 18 months. The opinion is forwarded by ECHA to the Commission for a final decision on the harmonized classification and labelling of the substance to be taken via comitology.

While the underlying database supporting a specific substance's assessment presented in the DAR and CLH proposal can be assumed to be broadly similar, the nature (i.e., the level of detail reported/presentation of study results) may differ as a result of the differing guidance and objectives of the two processes. The judgments made in relation to a particular piece of information may differ when considered under the hazard-based CLH process compared with the risk-based authorization process. Therefore, the DAR for approval and the CLH dossier for classification and labelling decisions may require different preparation and presentation of the underlying data, and not all data will be equally relevant for both decision-making procedures. To meet the regulatory objectives efficiently, both procedures require dossier formats specifically tailored to the different regulatory processes. The PPP Regulation requires a specific dossier structure and the CADDY electronic format system is used, whereas under the CLP Regulation there is a legal requirement for use of IUCLID (IUCLID 5 being the current version) which is a quite different electronic submission system using structured files.

A close linkage between these two processes is therefore highly desirable, especially for new active substances without existing, legally binding CLH in Annex VI of the CLP Regulation, or for active substances already classified which have to be re-evaluated in the light of new data that may necessitate revision of the existing classification and labelling.

The classification and labelling of active substances for human health endpoints is not only a principal criterion for the approval of active substances, safeners and synergists, but also the main basis for decisions on other regulatory categories and criteria established in the PPP Regulation namely:

---

[1] ECHA conclusions CLH Workshop 16 February 2011, ECHA

- consideration as low-risk active substances;
- identification as candidates for substitution;
- decisions on the interim criteria for endocrine disrupting properties that may cause adverse effects in humans;
- decisions on the relevance of metabolites that can occur in groundwater;
- decisions on toxicity with regard to defined persistence, bioaccumulation and toxicity (PBT) properties;
- setting risk mitigation measures for operators, workers, bystanders and residents in the procedure of national authorization of PPPs.

Therefore, a finalized harmonized C&L for active substances is, in many cases, a prerequisite for the harmonized authorization of PPP and mutual recognition according to the PPP Regulation. Furthermore, a final classification and labelling of the active substance is also essential for comparable decisions on approval of active substances in plant protection products under the PPP Regulation and biocidal products under the new Biocidal Products Regulation which will be published in 2012. Although the new Biocidal Products Regulation was not the subject of the workshop, part of the workshop results could have a positive impact on the classification procedure of biocides since the new Regulation will include cut-off criteria comparable to the PPP Regulation.

## 2. Workshop results

The main objectives of the workshop were to discuss options on how the two processes can most efficiently be aligned at the level of Member State authorities, EFSA and ECHA in the plenary session and in two main breakout group topics:

i. streamlining and integration of the review procedures for active substances in PPP for approval under the PPP Regulation and for classification and labelling under the CLP Regulation.
ii. scientific and practical issues in the interpretation of carcinogenicity, mutagenicity and reproductive toxicity studies and reporting regarding the criteria and practicalities in preparation of dossiers under both legislative frameworks.

The workshop started with a plenary session with lectures to introduce the two regulatory frameworks and provide technical information on the individual processes before entering into detailed discussions in two breakout groups. The presentations given in this first plenary session are available in Annex III of the workshop report which is published on the European Commission website:

(http://ec.europa.eu/food/plant/protection/evaluation/docs/report_berlin_april2011_en.pdf).

## 3. Streamlining and integration of the review procedures

Before the workshop, EFSA and ECHA had already started an exchange of information in order to identify practical solutions for processing proposals for CLH (especially for the

CMR hazard classes) concerning active substances in PPP quickly and efficiently and, as far as possible, within the same timeline as that of the risk assessment procedure. Based on a discussion paper prepared in February 2010 by ECHA on the cooperation between ECHA and EFSA in the assessment of hazard properties of active substances in PPP under the CLP and PPP Regulations, and a discussion at the meeting in June 2010 of EFSA's Network with the Member State authorities in the area of pesticides, the Pesticide Steering Committee, **the following scope was proposed as a starting point for the discussion in breakout group 1:**

- Streamlining and integration of the review procedures for active substances in PPP for approval under the PPP Regulation and for classification and labelling under the CLP Regulation.

The main goals of breakout group 1 were

1. to inform the discussion on **how the two processes could most efficiently be aligned** between Rapporteur Member States (RMS), EFSA and ECHA;
2. to consider the **anticipated workloads** stemming from the PPP active substance programmes in relation to the capacity of the **EFSA/ECHA process** with a view to ensuring that appropriate planning, management and prioritization procedures can be established;
3. to **raise awareness in Member States** (i.e., Competent Authorities (CAs) responsible for the evaluation of active substances in PPP and for their classification and labelling, respectively) and to communicate the importance of the issue and possible solutions;
4. to provide feedback on **a draft working document on processes** "Cooperation between CAs in Member States, ECHA and EFSA in the assessment of CMR properties of active substances in PPP under Regulations (EC) No 1107/2009 and 1272/2008" (based on the ECHA discussion paper from February 2011 regarding the preparation of the CLH report and the cooperation of the dossier submitter with RAC).

## 3.1. How the two processes could be aligned

Based on discussions held during the workshop, the following practical solutions were identified for new or existing active substances without existing legal C&L or for substances with legal C&L which have to be re-evaluated in consideration of new data for C&L:

- The Rapporteur Member State for the active substance should identify as early as possible in the evaluation process for approval or renewal of approval (preferably at the end of the completeness check) the need for an initiation of the CLH procedure under the CLP Regulation and should make a notification of intention for the CLH procedure to ECHA at an early stage. The notifier should be encouraged to indicate the classification and labelling in the PPP dossier.
- Specific issues, such as substance ID, for both approval under the PPP Regulation and inclusion under Annex VI of the CLP Regulation, should be solved as soon as possible by direct contact between the RMS and EFSA/ECHA.
- The RMS for the active substance could prepare in parallel:

- the DAR for EFSA for developing conclusions on possible fulfilment of the approval criteria to be sent to the Commission
- a proposal for harmonized classification and labelling for ECHA in accordance with the CLP Regulation, as well as ECHA's guidance and format requirements for developing a RAC opinion on classification and labelling
- Ideally, the CLH report should be ready and submitted one month before the DAR in order to allow time for the accordance check.

- EFSA and ECHA should aim to conduct their public consultations at the same time (EFSA for 60 days and ECHA for 45 days) to streamline the processes.
- The time schedule in EFSA for adopting the conclusions on fulfilment of the approval criteria is 120-150 days from the end of the commenting process, after which the Commission has 6 months for preparing its review report and a draft regulation.
- ECHA and EFSA will follow closely and potentially participate in the deliberations during each other's review process. To avoid duplication of work, leading actors of both processes will keep each other informed on the progress, identify critical issues as early as possible and, if necessary, organize joint discussions in dedicated ad hoc groups assembling capable experts for the issue under consideration from both processes.
- RAC will start the consideration for agreement on the opinion as early as possible Although RAC formally has 18 months for providing their opinion, the scheduled procedure should allow the adoption of the opinion on adequate classification well before expiry of the 6 month period in which the Commission develops its review report and draft regulation after receiving EFSA's conclusion on whether the active substance is expected to fulfil the approval criteria in the PPP Regulation.

The above mentioned parallel, and partly joint, processing of the proposals – conclusion on expected fulfilment of the approval criteria by EFSA and on harmonized classification and labelling by ECHA – would assure that the RAC's opinion on fulfilment of the classification criteria (in particular for the CMR hazard classes) is delivered in time for the Commission to develop its review report and draft regulation (i.e., within 6 months of receiving EFSA's conclusions).

## 3.2. Workloads from the PPP programmes in relation to the capacity of the ECHA

In order to ensure that any agreed aligned processes can deliver conclusions on harmonized C&L in an efficient and timely manner there was also a need to consider:

- the anticipated workloads stemming from the PPP active substance process and
- the capacity of the ECHA process to deliver conclusions taking into account available resources and other demands on those resources.

Proposals for harmonized classification of PPP active substances may be submitted from the following EFSA work programmes:

- new active substances: it is possible that a considerable number of new active substance/ safener and synergist applications will be submitted each year;

- renewal programme for existing active substances: this programme will start in 2013 (R2) and continue with substance assessments being delivered in 2015 (R3), and each year thereafter for the foreseeable future;
- review of safeners and synergists: likely to be low numbers of assessments submitted to EFSA from 2016 or beyond.

In addition, it is possible that limited numbers of requests may arise on an ad hoc basis as part of the Commission obligations to establish, by 14 December 2013, a list of (approved) substances that satisfy the criteria for candidates for substitution.

A proportion of the existing substances will already have harmonized classifications (i.e., mainly in the renewal programme for existing active substances). However, the use of hazard classification, as part of the criteria for approval and in relation to other areas (e.g., candidates for substitution/interim endocrine disruption criteria), may result in the generation of further studies to support updates/revision of existing proposals. The demands from the existing substance 'renewal' programme and priorities could be estimated at an early stage based on pre-submission information (updating statements). Initial information on priority for CLH consideration for new active substances could be gathered in the pre-submission process.

Therefore, consideration should be given to the establishment of agreed procedures for the management and prioritization of PPP active substances entering the process together with transparent procedures for monitoring their progress and the delivery of conclusions. The need for linkages between the annual planning and resource management processes within EFSA and ECHA should also be taken into account.

### 3.3. Raise awareness in Member States

The need for communication of the importance of a harmonized classification process in the approval process for PPP active substances was emphasized in the workshop. It was noted that it might be challenging to establish communication structures between the two processes at the national level due to the number of and coordination among involved governmental ministries and agencies.

However, the role of existing structures within the PPP assessment and decision-making processes in communication and raising awareness should be considered.

In particular, the roles and responsibilities of the following in communicating/planning/disseminating information should be considered, as well as the linkages between them:

- the Pesticide Steering Committee;
- the Standing Committee on the Food Chain and Animal Health;
- the Committee for Risk Assessment (RAC);
- the Competent Authorities for REACH and CLP (CARACAL, to advise the European Commission and ECHA on questions related to REACH and CLP).

There is a need to identify ways to facilitate continuous cooperation/scientific knowledge exchange at the national level among experts from different concerned authorities.

A critical element for ensuring a proper coordination is a full understanding of the different procedures according to the PPP and CLP Regulations and cooperation of the Member State acting as rapporteur under the EFSA process and dossier submitter under the ECHA process.

The RAC procedures are based on the full involvement of the dossier submitter, which does not end with the submission of the dossier. The dossier submitter is involved in the assessment of the comments received during the public consultation and should facilitate the RAC discussion by providing clarifications if needed.

CARACAL is composed of representatives from Member State competent authorities for REACH and CLP, representatives from competent authorities of EEA-EFTA countries, as well as a number of observers from non-EU countries, international organizations and stakeholders. The EUROPEAN COMMISSION (DG Enterprise and Industry and DG Environment) will prepare a proposal to adapt the CLH in Annex VI to the CLP Regulation to technical progress every year based on the opinions received from ECHA's RAC for harmonized classification and labelling.

ECHA is currently updating the process and cooperation between RAC and the MS as dossier submitters based on the outcome of the workshop "On the Way to CLH" held in February 2011.

The discussions at this workshop covered issues and procedural changes such as:

- changes in the Registry of Intentions;
- accordance check streamlining;
- facilitation of communication between dossier submitters, ECHA and RAC;
- dealing with comments received during public consultation;
- withdrawal and resubmission procedures in the case of receipt of new crucial information during public consultation or even at a later stage.

The Member State acting as rapporteur under the EFSA process and dossier submitter under the ECHA process should be fully familiar with the RAC process. ECHA will provide information if required. A full internal coordination among the MS experts and CAs is particularly essential when there is more than one CA involved in the process.

## 3.4. Finalize a draft working document on processes

During the workshop the working procedures were discussed intensively. The outcomes of the discussions are reported in the presentations as included in the report published on the Commission website: **http://ec.europa.eu/food/plant/protection/evaluation/docs/report_berlin_april2011_en.pdf.**

The workshop did not conclude on a draft working procedure, however, to keep up the momentum, the workshop Organizing Committee took the initiative to develop a draft working document on processes which will serve as a basis for the first projects in the parallel processing of dossiers and which has been presented to Member States' competent authorities.

## 4. Scientific and practical issues in the interpretation of studies and reporting

Based on the criteria for the approval of active substances in PPP under the PPP Regulation and the classification criteria regarding "Health hazards" under the CLP Regulation, the following scope was proposed as a starting point for the discussion in breakout group 2:

- scientific and practical issues in the assessment and interpretation of carcinogenicity, mutagenicity and reproductive toxicity (CMR) studies, and requirements concerning adequate preparation of dossiers (with respect to scientific content and formatting according to the PPP Regulation and the CLP Regulation).

The main goals of breakout group 2 were:

1. to **recommend solutions regarding formatting problems** with documents/dossiers (e.g., how to facilitate compilation of CLH dossiers by the Rapporteur Member Stat, how to integrate additional relevant documents from the pesticide process in these dossiers, possibility of profits for CLH dossiers based on experience with previous pesticide assessments);
2. to discuss possibilities and **practicalities for submission of IUCLID 5 dossiers** in addition to the dossiers for active substances under the PPP Regulation to facilitate the preparation of dossiers for classification and labelling, as well as possible assistance for approval;
3. to **improve harmonized interpretation and reporting** of carcinogenicity, mutagenicity and reproductive toxicity studies, discuss scientific principles of interpretation of relevant studies. This shall contribute to avoiding conflicting interpretations and different reporting of the same studies under the two processes.

### 4.1. Recommended solutions regarding formatting problems

The workshop participants recognized that although in the current DARs the purpose of the substance evaluation is mainly to derive a basis for risk assessment (i.e., deriving NOAELs/LOAELs and setting reference doses) this issue requires reconsideration due to the new cut-off criteria settled in the PPP Regulation. The main intention of the CLH report is hazard identification (i.e., assessment of the nature and severity of effects and the dose response relationship to be compared with a defined set of criteria) including the specific comparison of the available evidence with the CLP classification criteria.

Currently the structure of the DAR is under discussion and will be revised in the next few years. A proposal for this revision was presented the break out group session.

For CMR substances, the DAR under the PPP Regulation would require a similar assessment (hazard identification and comparison with the criteria) to that required for the

CLP report, and therefore, the same document can cover both assessments. For other hazard classes, the DAR should also be the basis for the CLH proposal, and therefore, it seems logical to integrate this information as well.

The workshop participants considered that for a better common scientific understanding, it is essential to implement the same structure in the reporting and formatting of the DAR and CLH reports. In fact, the proposed solution is to incorporate the weight of evidence and comparison with the CLP criteria to be included in the CLH report as one of the chapters/documents/elements of the new DAR structure. Additional considerations are needed for facilitating the description of the key studies results in a way that could cover the needs for the DAR and for the CLH report. The structure of the CLH report is defined in the legislation (reference to Chemical Safety Assessment and Report under REACH) and described further in the CLP guidance, which allows the required flexibility to accommodate the dossier's specific needs. It should be kept in mind that the CLH process also applies to biocides and industrial chemicals, that some substances have several uses and that the CLH structure must be similar for all types of chemicals. However, as the structure and level of detail of the CLH report will be periodically updated based on RAC experience when processing the CLH dossiers, the specific input gained during the discussions of the new DAR format can be used in the periodic revisions of the CLH report format. In addition, specific guidance for preparing the CLH report as the hazard identification chapter of the DAR for PPP active substances is required.

As an outcome of the ECHA workshop "On the way to CLH", RAC, with the support of the ECHA secretariat, is currently revising the structure of its opinions and particularly of the background document presenting the detailed justification of the RAC opinion. The background document is based on the original CLH report. On the other hand, the PPP experts are currently discussing possible improvements to the structure of the DAR and dossier. It was considered that ECHA and EFSA should be in close contact during these developments in order to ensure mutual feedback and coordination between both processes.

To complement the proposal mentioned above, it was also recommended that when drafting the DAR annexes related to the robust study summaries and the assessment summaries which constitute the basis for the hazard identification and risk assessment, both intentions should be kept in mind and addressed, allowing the use of the text related to the hazard identification as the starting point for the CLH report and DAR hazard identification chapter.

It was also mentioned that currently some DARs do not contain a proper presentation of the evidence related to the hazard identification and its comparison with the CLP classification criteria. It was highlighted that this is an essential part of the CLH report and should be specifically considered. The current RAC experience might offer further suggestions for reporting the weight of evidence and the comparison of data with the criteria, and some examples were presented during the workshop.

## 4.2. Practicalities for submission of pesticide dossiers in IUCLID format

The OECD Expert Group on the Electronic Exchange of Pesticide Data makes an effort to support the harmonization of the international submission formats used for pesticide registration (Caddy, eIndex, ePRISM). This harmonized format is called GHSTS (Global Harmonized Submission Transport Schema) and will be finalized in 2012. At present it is not possible to submit a full document-based pesticide dossier from a company to the authority using IUCLID 5, which is endpoint record-based. The answer should be found by ECHA by evaluating the proposals collected in a public consultation.

- The objective of this public consultation, organized in collaboration between OECD and the ECHA, is to receive input and exchange ideas on the next generation of the IUCLID software from stakeholders not represented at the OECD IUCLID User Group Expert Panel.
- After a few years of experience in using IUCLID 5 as a tool for collecting, storing and exchanging information on chemicals in the OECD, and for national and regional chemical review programmes, it is time to plan for the next generation of the IUCLID software and to adapt it to fit the evolving needs of a growing user community. Example developments could be the extension of IUCLID to specific information relevant for pesticides or information on exposure and risks related to uses of substances, or the development of several user interfaces adapted for a specific purpose connected to the same core database.

This next generation of the IUCLID software might also be useful for the submission of a future PPP dossier and/or a DAR, as well as the CLH report. The Harmonized Templates were implemented to store structured data from studies on an endpoint record level. This technique is used in IUCLID.

- The content of the XML files according to the Harmonized Templates shall replace the Tier II summary level (Word, PDF) to prevent a duplicate lifecycle management by the companies of a text and of the corresponding structured data set.
- Today the authorities have to produce a duplicate lifecycle management of a CLH text and a technical CLH dataset in parallel over a long period. Why is it necessary to produce two versions of a CLH dossier, a text processor CLH dossier and the technical IUCLID data file?

A mutual understanding of the needs and workload implications was the starting point for this discussion. There is a clear benefit in having an IUCLID 5 dossier for all substances, including PPP active substances, but on the other hand there is an additional workload for the CAs when preparing an IUCLID 5 dossier from a dossier presented in a different format. Over the long-term the OECD approach may provide a fully compatible solution and this was recognized as the best solution.

The workshop participants recognized that the role of the PPP CAs should be equivalent to the role of the REACH/CLP and biocide CAs: to revise and update the IUCLID 5 dossier presented by the relevant companies. Therefore, before a fully compatible submission system is developed, the alternative should be to request the companies to include in their submission an IUCLID 5 dossier for the studies relevant for classification and labelling.

## 4.3. Improve harmonized interpretation and reporting

The ECHA and EFSA processes represent the scientific assessments of the available information in order to establish solid scientifically based conclusions for supporting the decision-making process by the European Commission. ECHA and EFSA have specific mandates, defined in their respective regulations. The workshop discussions and the conclusions presented below should be understood and implemented taking into account the different and independent roles and mandates of ECHA and EFSA, and the European Commission.

The conclusion that the CMR-related **cut-off criteria** for active substances to be included in PPP are met is based on a conclusive scientific assessment on the substance with regard to the fulfilment of the approval criteria proposed by EFSA[2] and on the opinion of ECHA. In order to support such a conclusion early in the evaluation process under the PPP Regulation, common interpretation of the classification criteria for CMR properties in both contexts (EFSA and ECHA) would be an important prerequisite. Both agencies should cooperate to achieve a common interpretation of the underlying studies, particularly in terms of reliability and relevance, and to explain any divergence and deviation if needed.

Classification as CMR category 1A or 1B will exclude an active substance from approval and subsequent use in PPPs (unless exposure is negligible in case of CR), whereas classification as CMR category 2 allows approval. The credibility of the scientific assessments of CMR properties could suffer if conclusions under the PPP Regulation and under the CLP Regulation were inconsistent, e.g.:

- if a CLP decision adopted by the Commission on the basis of a RAC opinion (CMR category 1A or 1B) made it necessary to revoke an active substance approval, which was adopted at an earlier time point or
- if active substance approvals were declined earlier in the process on the grounds of an RMS or EFSA proposal for CMR category 1A or 1B classification, but later a CLP decision adopted by the Commission on the basis of the RAC opinion resulted in CMR category 2 classification which would have allowed the approval of the active substance.

Similarly, divergences in the answer to the question of whether a substance should be classified as CMR category 2 or should not be classified would also have consequences at PPP authorization level and even at active substance approval level (relevance of groundwater metabolites). Harmonized application of the CLP criteria for CMR classification within the EU Member States and by different Expert Meetings is therefore essential.

---

[2] Article 12(2) of Regulation 1107/2009 lays down that "...the Authority shall adopt a conclusion in the light of current scientific and technical knowledge using guidance documents available at the time of application on whether the active substance can be expected to meet the approval criteria provided for in Article 4... "

Although detailed criteria for hazard classification and labelling of substances have been laid down under the CLP Regulation, particularly the specific criteria for CMR classification requires expert judgement and consideration of many different factors (e.g., weight and strength of evidence, mechanism or mode of action and its relevance to humans) included in the available relevant experimental data and the additional reliable information.

Based on the experience from various national and international Expert Meetings, it seems obvious that the interpretation of CMR data from experimental tests and epidemiological studies by different Expert Meetings in ECHA and EFSA, i.e., the RAC and the Pesticide Risk Assessment Peer Review (PRAPeR) meeting, does not necessarily lead to the same opinion and proposal on classification, even though the same data have been evaluated. The current RAC experience already indicates a significant number of borderline cases as being particularly problematic. The workshop participants considered that the optimal solution would be an involvement of the experts at an early stage. This requires coordination within the rapporteur MS under the PPP process, as well as between EFSA and ECHA. Ideally, the RAC opinion on harmonized classification and labelling should be the basis for the EFSA conclusion on the cut-off criteria related to CMR properties; if this is not feasible in all cases, the RAC opinion on the CMR classification should be at least available for the Commission for their decision-making process on the approval. There is a special need for a common interpretation of the criteria for the classification of substances for reproductive toxicity (paternal and maternal toxicity, consideration of potency and setting of specific concentration limits, developmental versus lactation effects, etc.).

When expert judgement and consideration of many different factors (e.g., weight and strength of evidence, mechanism or mode of action and its relevance to humans) is needed, common scientific understanding is essential under both regulations.

The workshop participants considered that the cooperation of the experts involved in both processes is essential and encouraged ECHA and EFSA to consider this need when establishing their processes. The ideal solution, particularly for borderline cases, would be to organize a single detailed expert discussion that could feed into both processes. The working procedures from RAC and EFSA already allow the participation of invited experts and a set of consultations with the committees. ECHA and EFSA were requested to coordinate the involvement of the relevant experts, ensuring that all relevant information is available to the experts, and to establish mechanisms for facilitating the exchange of views among the experts early in the process for identifying divergent interpretations, and organize ad hoc expert discussion platforms in order to try to get consensus on the scientific interpretation of the data.

The rapporteur MS under the PPP Regulation, acting as dossier submitter for the CLH dossier, plays a key role in both processes. It is essential that when reporting the studies' results, weight of evidence and its comparison with the CLP criteria, the MS experts consider specifically the RAC needs and previous opinions on similar cases. Following the RAC decision, the RAC Manual of Conclusions and Recommendations will be available to the CAs in order to facilitate this process.

## 5. Main conclusions and recommendations

This chapter includes a summary of the main conclusions agreed upon by the different breakout groups and a table presenting the main recommendations for actions to follow up.

It has to be underlined that the information below refers to those conclusions and recommendations made most frequently by the experts in the breakout groups' sessions and in plenary.

The main conclusions of the workshop can be summarized as follows:

- need to inform ECHA as early as possible on a potential candidate for CLH classification;
- call for prioritization of proposals for harmonized classification and labelling suggesting classification as CMR;
- the importance of increased cooperation and awareness among the different competent authorities;
- ensure consistency with respect to information evaluated under both processes and harmonization of the currently different formats for hazard assessment;
- progress toward a harmonized electronic system for submission of data;
- the relevance of the proper presentation of evidence related to hazard identification and comparison with CLP criteria for harmonized interpretation on CMR studies;
- harmonized reporting on CMR studies and integration into the current reporting formats under the two processes.

The workshop concluded that in the long-term "one substance, one dossier, one procedure and one discussion" would be the ideal situation.

## Disclaimer

This document is based on the workshop report as published in the SANCO webpage which was elaborated upon by members of the Organizing Committee of the Workshop including members of the European Commission DG Health and Consumers, of the European Chemicals Agency (ECHA), of the European Food Safety Authority (EFSA) and Member States' representatives. It does not necessarily reflect the views of the Commission Services, ECHA services or Member State agencies, but it reports the discussed topics and outcomes of the workshop. The authors and the representatives mentioned in the Acknowledgement Section were all members of the Organizing Committee.

## Author details

Roland Solecki and Vera Ritz
*Federal Institute for Risk Assessment, Germany*

Abdelkarim Abdellaue
*Norwegian Food Safety Authority, Norway*

Teresa Borges
*Committee for Risk Assessment, European Chemicals Agency*

Kaija Kallio-Mannila
*Safety and Chemicals Agency, Finland*

Herbert Köpp
*Federal Office of Consumer Protection and Food Safety, Germany*

Thierry Mercier
*French Agency for Food, Environmental and Occupational Health and Safety, France*

Gabriele Schöning and José Tarazona
*European Chemicals Agency*

Cooperation at European level in the assessment of human health hazards of active substances in Plant Protection Products under Regulation (EC) No 1107/2009 and the harmonized classification and labelling of active substances under Regulation (EC) No 1272/2008.[3]

## Acknowledgement

The authors would like to thank Steve Dobson, Herman Fontier, Patrizia Pitton and Anniek van Haelst for their essential contributions to the work in the Organizing Committee, as well as their valuable comments and input to the manuscript. They would also like to thank the participants of the workshop held in Berlin in April 2011 for their fruitful discussions and input at the workshop, and during the preparation of the workshop report (http://ec.europa.eu/food/plant/protection/evaluation/docs/report_berlin_april2011_en.pdf.).

## 6. References

[1] Regulation (EC) No 1107/2009 of 21 October 2009 concerning the placing of plant protection products on the market and repealing Council Directive 79/117/EEC and 91/414/EEC. Official Journal of the European Union, 24.11.2009., L309/1-L309/50, http://eurlex.europa.eu/LexUriServ/LexUriServ.do?uri=OJ:L:2009:309:0001:0050:EN:PDF

[2] Regulation (EC) No 1272/2008 of 16 December 2008 on classification, labelling and packaging of substances and mixtures, amending and repealing Directives 57/548 (EEC and 1999/45/EC and amending Regulation (EC) No 1907/2006. Official Journal of the European Union, 31.12.2008., L353/1-L353/1355, http://eur-lex.europa.eu/LexUriServ/LexUriServ.do?uri=OJ:L:2008:353:0001:1355:en:PDF

[3] OJ L 309, 24.11.2009
OJ L 353, 31.12.2008

Section 2

# Environmental and Stress Plant Physiology and Behavior

Chapter 2

# Grain Yield Determination and Resource Use Efficiency in Maize Hybrids Released in Different Decades

Laura Echarte, Lujan Nagore, Javier Di Matteo, Matías Cambareri, Mariana Robles and Aída Della Maggiora

Additional information is available at the end of the chapter

http://dx.doi.org/10.5772/55287

## 1. Introduction

Maize (Zea mays L.) grain yield have increased during the last decades. A recent review [1] indicated genetic grain yield gains of 74 to 123 kg $ha^{-1}$ $year^{-1}$ for different time periods between 1930 and 2001, in the US corn belt, Argentina and Brazil [2-6]. Current reviews on the physiological processes associated with those yield increments have been focused on US corn belt hybrids and maize hybrids of Ontario, Canada [e.g. 1; 7; 8]. As such, grain yield increments were associated mainly with an increased kernel number, a consistently improved stay green, and a longer period of grain fill. Those reviews agreed on that harvest index (HI; i.e. the relationship between grain yield and final shoot biomass) did not consistently change over time; in contrast, HI of Argentinean maize hybrids have increased during the 1960-1990 period [9; 10]. This review will be focused on the ecophysiological mechanisms contributing to the greater yield in modern than in older maize hybrids; with particular interest in Argentinean maize hybrids because they have shown a distinctive trait change over the years (i.e. HI increment).

### Grain yield

Grain yield can be expressed as the product between shoot biomass and harvest index. In Argentina, harvest index was increased while shoot biomass was not consistently increased over the years during the period 1965-1993 [11]. As such, HI increased from 0.41 to 0.52 in maize crops growing under optimal conditions [9]. The increased harvest index was associated mainly to a greater increase in grain yield numerical components (i.e. kernel number and/or kernel weight) than in shoot biomass. On the contrary, shoot biomass has increased while harvest index have remained constant in maize hybrids released in Canada and the US in different decades [1; 12]. Most of the shoot biomass accumulation increments in those hybrids, occurred during the grain-filling period [13; 14]; and they were mainly associated with an increased capacity of maintaining higher leaf photosynthetic rate of green leaf area (i.e., functional "stay green") during the grain-filling period [15-17]. The next sections will review the main processes influencing grain yield numerical components determination (i.e. kernel number and kernel weight) and their changes in Argentinean maize hybrids released in different decades. Implications on stress tolerance and resource use efficiency will be also discussed.

**Kernel number**

Kernel number is the main yield component accounting for grain yield increments over the years [18; 19]. Figure 1 illustrates a conceptual framework of the main processes contributing to kernel number determination in maize.

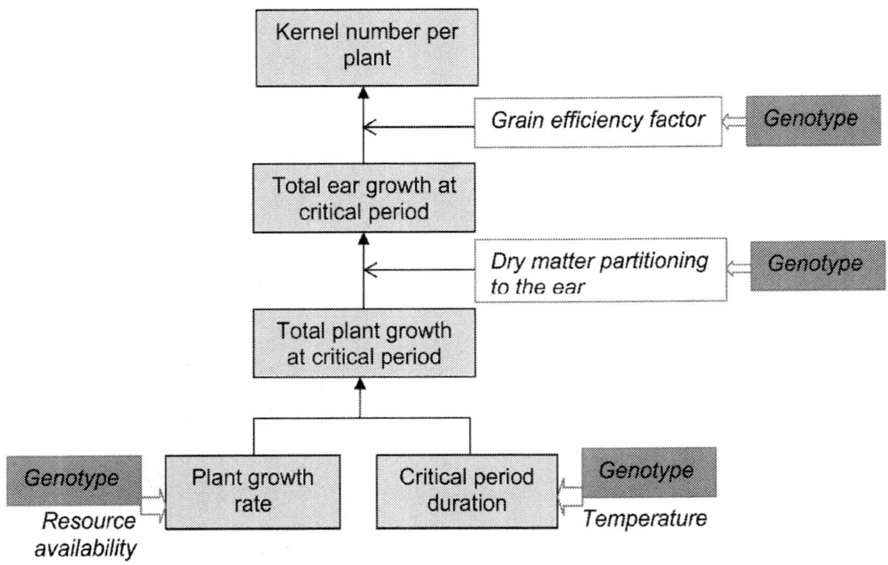


**Figure 1.** General model for kernel number determination in maize (Adapted from Andrade et al. (20)).

Kernel number per plant is a function of the physiological condition of the crop or plant at a period of 15 days bracketing silking (i.e. critical period for kernel number determination; 21-26) or between -227 and 100°C day from silking [27]. As such, kernel number is a function of photosynthesis at silking [22] and it is closely related with plant growth rate during the critical period for kernel set [18; 28]. The relationship between kernel number per plant (KNP) and plant growth rate during the critical period for kernel set (PGRs) was described by two successive curves to account for the first and second ear in prolific hybrids, or a single curve in non-prolific hybrids [18; 28; 29]. A particular feature of the KNP-PGRs relationship is the significant PGRs threshold for kernel set that results in abrupt reductions in kernel number at low resource availability per plant [29]; which might reflect a strong apical dominance [24; 30]. Using contrasting plant densities along with individuals instead of plot means provide a wide range of values for PGRs and KNP; and it is possible to obtain more precise estimations of the threshold PGRs for kernel set [28; 29]. Allometric models are fitted to the relationship between shoot biomass and morphometric measurements (i.e. stem diameter, ear length, ear diameter) and are used to estimate the growth during the critical period for kernel set of individuals that remains in the field from sowing to physiological maturity (i.e. individual plant methodology, 29). The regression between estimated shoot biomass using allometric models and the actual shoot biomass of plants before silking is depicted in Figure 2 and it shows an example of the reliability of the individual plant methodology.

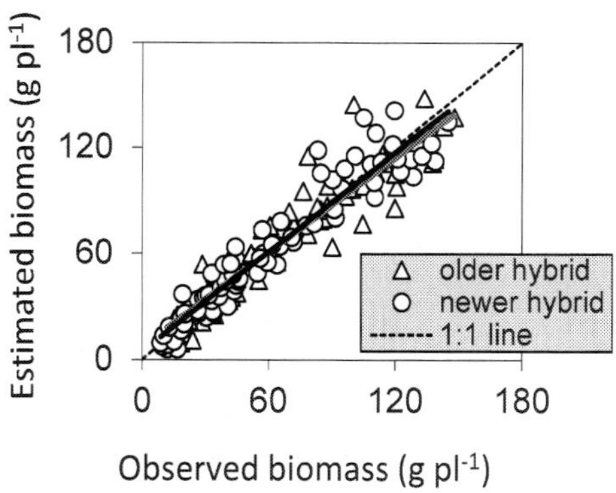


**Figure 2.** Relationship between estimated and actual shoot biomass at the beginning of the critical period for kernel set, for an older (DKF880) and a newer (DK752) maize hybrid. Shoot biomass was estimated using allometric models. The dotted line shows the 1:1 ratio and the solid lines show the fitted model for the older (gray) and the newer (black) maize hybrids. Fitted linear equations were $y = 0.89\,x + 7.1$, $R^2 = 0.87$, n=71 for the older hybrid, and $y = 0.94\,x + 4.2$, $R^2 = 0.91$, n=72 for the newer hybrid (Adapted from Echarte et al. (10)).

A comparison of the KNP-PGRs relationship among 5 Argentinean hybrids released between 1965 and 1993 established that newer hybrids set more kernels per unit PGRs than older hybrids as was indicated by (i) the lower threshold PGRs for kernel set and (ii) the greater potential kernel number at high availability of resources per plant, for newer than for older hybrids (10; Figure 3). Plant growth rate during the critical period for kernel set at each plant density did not show a clear trend with the year of release. The lower threshold PGRs for kernel set contributed to reduce the number of sterile plants in modern than in older maize hybrids and thus to a higher kernel number per plant as resource availability per plant decreases. Other authors also found less % of barren plants in newer than in older hybrids [31; 32; 18]. The lower threshold PGRs for kernel set could have probably resulted from indirect selection of genotypes under progressively higher plant densities and from a wide testing area that includes low-yield environments [33-38]. The determination of the thresholds of plant growth rate for kernel set were recently suggested as a phenotyping trait in breeding programs (39). However, the individual plant methodology [29; 10] seems more suitable for a reliable estimation of PGRs thresholds for kernel set than the mean PGRs per plot calculated in other works [18; 40]. At high resource availability per plant, the greater potential kernel number in the topmost ear contributed to a high KNP [9; 10]. Although differences were found among hybrids, there was not a clear trend with the year of hybrid release in threshold PGRs for prolificacy, nor in percentage of prolific plants beyond that threshold [9]. Also, no significant changes in ears per plant for US maize genotypes released between 1930 and 1980 were evident [32]. However, an increase in prolificacy with the year of hybrid release was reported in other works [3; 18].

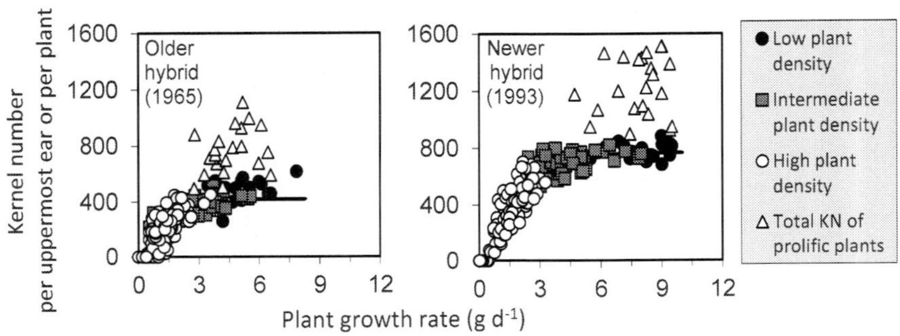


**Figure 3.** Relationship between kernel number per uppermost ear or per plant and plant growth rate during a period bracketing silking (PGRs) in an older (DKF880) and a newer (DK752) maize hybrid released in Argentina in different decades (year of release between brackets). Triangles represent kernel number of prolific plants (kernel number of the topmost plus the second ear). Other symbols represent KN of the topmost ear at low (2-4 plants $m^{-2}$; solid circles); intermediate (8 plants $m^{-2}$; squares), and high plant densities (16-30 plants $m^{-2}$; white circles). Adapted from Echarte et al. (10).

A greater dry matter partitioning to the ear (i.e. ear growth rate per unit PGRs) and/or a greater grain efficiency factor (i.e. kernel set per unit of ear growth rate during the critical period for kernel set) are physiological processes contributing to a greater KNP per unit PGRs (41; Figure 1). It has been stated that kernel set improvements with the year of the hybrid release were attributable to (i) increased partitioning of dry matter to the ear during the critical period for kernel set at low and intermediate resource availability per plant; and to (ii) greater kernel set per unit of ear growth rate at high resource availability per plant (10; Figure 4). Previous works have shown dry matter partitioning to the ear increments as a result of a reduction in tassel size or tassel removal [24; 42; 43]. Greater dry matter partitioning to the ear in newer compared with older maize hybrids is in agreement with the declined tassel size of US hybrids from the 1930s to the 1990s [15]. Tassel branch number and dry weight were reduced over the years in US hybrids [2; 3]. At high resource availability, the greater kernel set per unit ear growth rate was mainly attributable to the greater potential kernel number per ear [10]. Other processes contributing to elucidate differences among hybrids in grain efficiency factor, like a lower assimilate requirement per kernel [21; 40; 42] or a more synchronous fertilization of florets within the ear [44;45], did not show a clear trend with the year of the hybrid release [10]. The inherent greater stand uniformity of the single-cross modern than in double-cross older hybrids was not an additional factor influencing kernel set per unit PGRs; since, plant size variability at the critical period for kernel set was similar among hybrids of different decades [9].

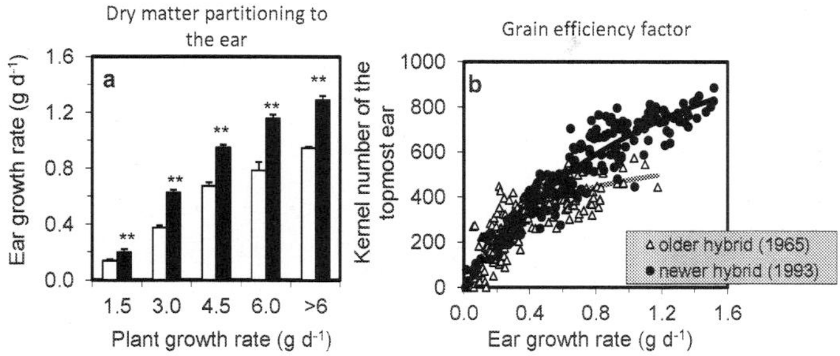


**Figure 4.** Dry matter partitioning to the ear (a) and grain efficiency factor (b), for an older (DKF880) and a newer (DK752) maize hybrid released in Argentina in different decades (year of release between brackets). Bars indicate standard error. ** indicates significant differences between hybrids at P< 0.05. Adapted from Echarte et al. [10].

The modifications to the features of the relationship between KNP and PGRs (i.e. lower threshold PGRs for kernel set and greater potential kernel number) were associated with a more uniform HI across resource availabilities in newer than in older maize hybrids (Figure 5; 9). At low resource availability, decreases in HI were sharper in older hybrids. At high resource availability per plant, decreases in HI of non-prolific plants were less pronounced in newer than in older hybrids (Figure 5; 9);

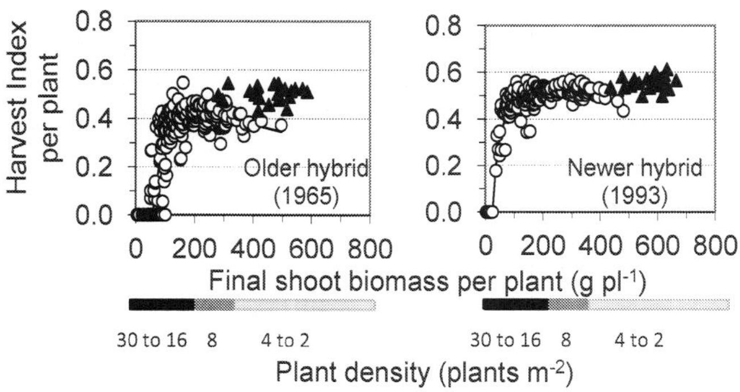


**Figure 5.** Relationship between harvest index per plant and final shoot biomass per plant in an older (DKF880) and a newer (DK752) maize hybrid released in Argentina in different decades (year of release between brackets). Triangles represent harvest index of prolific plants. Bottom bars represent the plant densities used to obtain the corresponding ranges of shoot biomass per plant. Adapted from Echarte and Andrade (9).

The lower threshold PGRs for kernel set was associated also with an improved tolerance to high plant density in newer maize hybrids [10]. Greater tolerance to high plant density was reported for hybrids released during different decades in the US, Canada and Argentina [3; 18; 32; 36; 40; 46]. The response of grain yield to plant density was curvilinear in Argentinean maize hybrids released between 1965 and 1993 [19] and between 1965 and 1997 [40], in agreement with the generally reported grain yield response to plant density for maize [47; 48]. Grain yield response to plant density was mostly associated with number of kernels per unit area [19], in accordance with other works [18; 47; 49]. In general, differences in kernel number $m^{-2}$ among hybrids released in different decades increased with plant density [19]. Figure 6 shows that kernel number $m^{-2}$ of a hybrid released in 1965 increased with plant density up to 8 pl $m^{-2}$; whereas, kernel number of a newer hybrid released in 1993 increased with plant density up to 14.5 plants $m^{-2}$. A recent study demonstrated that kernel number of current Argentinean maize hybrids (i.e. released in 2010) is consistently higher than that of an hybrid released in 1993 at high plant densities [50]. Greater tolerance to other stresses like weed competition (51), low night temperatures [16; 52], low soil nitrogen [17; 53; 54] and drought [55] were reported for hybrids released during different decades in the US and Canada. It was demonstrated that the nature of the environmental stress (e.g., plant density, nitrogen, water) causing variations in PGRs did not influence the KNP-PGRs relationship [56; 57]. Therefore, it is likely that a lower threshold PGRs is the underlying feature contributing to explain the greater general stress tolerance in newer than in older maize hybrids.

The greater kernel number at low plant density in newer compared with older maize hybrids (Figure 6) is another distinctive trait improved in Argentinean maize hybrids; since no grain yield improvement at very low plant densities was reported for US and Canadian hybrids [3; 37]. Moreover, although newer Argentinean hybrids released in 2010 yielded more than hybrids released in 1993 in a range of plant densities between 5 to 14.5 plants $m^{-2}$, the greatest grain yield improvement during the 1993-2010 period occurred at the lowest plant density (i.e. 5 plants $m^{-2}$; 49).

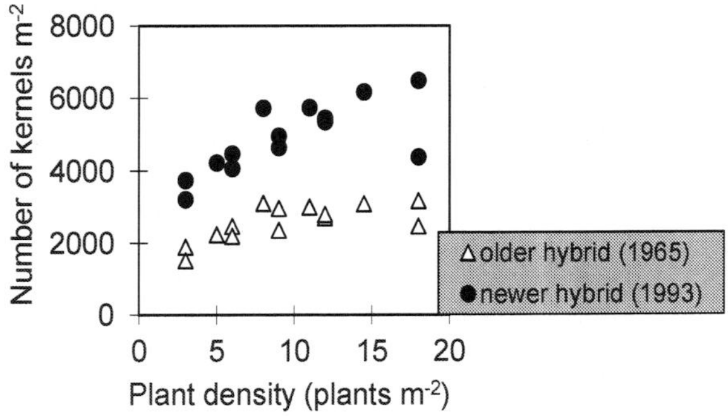


**Figure 6.** Number of kernels $m^{-2}$ as a function of plant density for an older (DKF880) and a newer (DK752) maize hybrid released in Argentina (year of release between brackets). Adapted from Echarte et al. (19).

**Kernel weight and chemical quality**

A general model for kernel weight determination in maize is shown in Figure 7. Although kernel weight differed among hybrids it did not show a clear trend with the year of hybrid release [60].

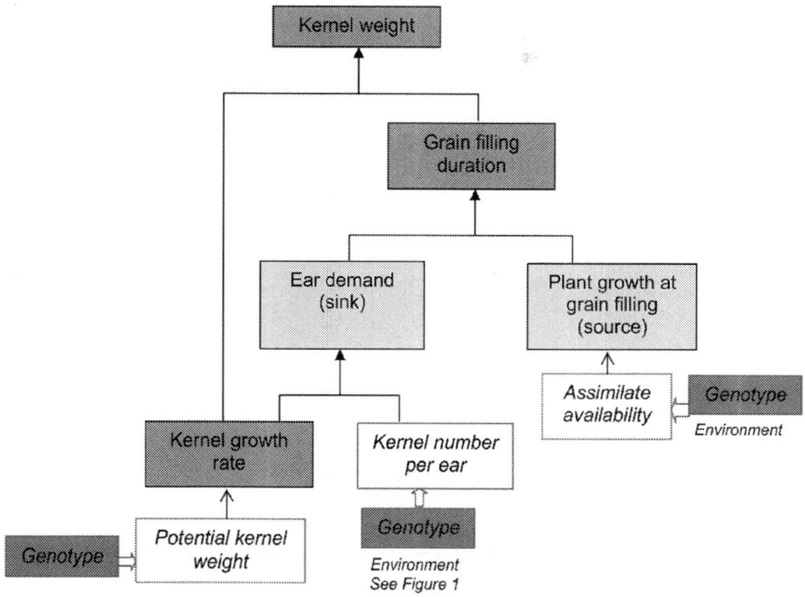


**Figure 7.** General model for kernel weight determination in maize.

Biomass accumulation in kernels begins shortly after fertilisation and it can be represented by a sigmoidal pattern in which a lag and a linear growth phase can be distinguished [58; 59]. Of the two components that determine final kernel weight (i.e. the kernel growth rate during the linear phase or effective grain filling period and the effective grain filling duration; Figure 7), kernel growth rate was the main component contributing to explain differences in kernel weight among hybrids released in different decades up to 1993 [60]. Kernel growth rate is strongly correlated with number of endospermatic cells and starch granules, which in turn determine the potential kernel size [61-63]. This contention suggests underlying differences among hybrids in potential kernel weight.

Duration of the grain filling period, and in turn kernel weight, is affected by the ratio between assimilate availability (source) and the potential capacity of the ear to use the available assimilates (i.e. ear demand, sink) during the grain filling period (Figure 7; 29; 65; 67-71). Since under optimal growing conditions, hybrids differ in kernel number per plant but also in kernel growth rate or potential kernel weight [60]; the ear demand (i.e. sink) was better described by both, the number of kernels per ear and their potential kernel weight (i.e., ear demand = KNP x kernel growth rate) rather than by KNP alone as in previous works [64-68]. As such, ear demand was greater in newer than in older hybrids by means of a greater kernel number per plant or a large potential kernel weight [60].There was not a clear trend with the year of the hybrid release in source-sink ratio in non-limiting environments (i.e. optimum resources availability; 59). An enhanced source-sink ratio (i.e. calculating the sink as kernel number alone) has been indicated for Argentinean maize hybrids released between 1965 and 1997 [40]. However, kernel weight reductions in response to source reductions due to defoliation during grain filling were greater in newer than in older hybrids (Figure 8a; 60). This response was associated with the greater ear demand relative to the source capacity in newer Argentinean maize hybrids (Figure 8b). Thus, if breeding for high yield potential continue increasing the ear demand without a proportional increment in total source capacity, kernel weight would be source limited and it will be more affected by source variations during the grain filling period in the newer maize hybrids. In agreement, ear demand of current Argentinean maize hybrids (i.e. released in 2010) was greater than that of maize hybrids released in 1993 [72]. As such, ear demand increased at a rate of 1.13% year$^{-1}$ during the last 45 years in Argentina; and kernel number was the main component influencing this increment rather than kernel growth rate [72]. In contrast, source-sink ratios were greater for newer than for older Ontario maize hybrids for the 1959-2007 period [8]. The increased functional "stay green" (i.e. capacity of a leaf to retain its photosynthetic rate during the grain filling period; 8) was the main factor underlying the larger source during the grain filling period in newer maize hybrids of the US corn belt and Ontario, Canada [1; 8; 17].

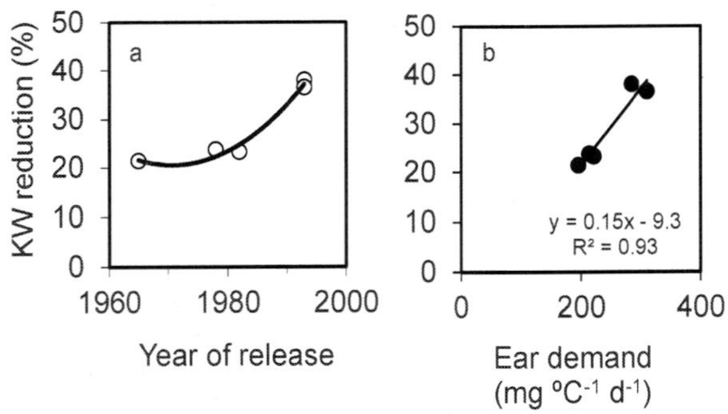


**Figure 8.** Kernel weight reduction (%) due to full defoliation during the grain filling period as a function of (a) year of hybrid release and (b) ear demand (mg $^{\circ}C^{-1}$ $d^{-1}$) for 5 hybrids released in Argentina from 1965 to 1993. Adapted from Echarte et al. [60].

The greater ear demand along with the genotypes used in the Argentinean maize breeding programs influenced the grain chemical quality of hybrids released in different decades [11]. Protein concentration decreased with the year of the hybrid release in an environment without nitrogen (N) fertilization but it was not modified when N was applied (Table 1); soil N-$NO_3$ level at V6 stage in this experiment was higher than the minimum required for maximum yield achievement (i.e. 27 ppm in this experiment versus a threshold of 24 ppm

$N\text{-}NO_3$ for maximum yield; 73). Protein concentration was negatively correlated with grain yield ($r=-0.79$, $p=0.06$) in agreement with previous findings [74-76]. The decline in protein concentration in kernels might have been the result of non-proportional increments of N and carbon fluxes to the kernels over the years. In addition, lower protein concentration in kernels were associated with low source-sink ratios [65; 77]. Similar trends in protein concentration over the years were reported for other crops [78; 79] and for US maize hybrids released during the period 1930-1991 [3]. On the contrary, protein concentration in kernels increased in Canadian hybrids released in different decades [80]. The increment in both, grain yield and protein concentration, might be associated with the increased source-sink ratio in Canadian maize hybrids [54]. As well, similar protein concentration under high N availability in Argentinean maize hybrids released in different decades might have been related to N luxury consumption [81; 82]. Oil kernel concentration was stable in hybrids released between 1965 and 1984; but it was reduced in hybrids released in 1993 ($r^2=0.84$, $p<0.05$, Figure 9). Oil is mainly located in the embryo [83; 84] and it is probable that the embryo-endosperm ratio has decreased with the year of the hybrid release. In agreement, embryo-endosperm ratio was greater in US hybrids selected for high oil concentration [85].

| Hybrid | Year of release | Protein ($g\ kg^{-1}$) | |
|---|---|---|---|
| | | No N fertilized | N fertilized |
| DKF880 | 1965 | 95.0 a | 98.7 a |
| M400 | 1978 | 93.0 ab | 95.7 a |
| DK4F36 | 1982 | 81.0 bc | 86.7 a |
| DK4F37 | 1985 | 77.7 c | 88.3 a |
| DK664 | 1993 | 78.0 c | 86.7 a |
| DK752 | 1993 | 79.7 c | 93.0 a |
| SE within columns | | 4.04 | |
| SE within rows | | 3.85 | |

**Table 1.** Protein concentration ($g\ kg^{-1}$) in grains of Argentinean maize hybrids released between 1965 and 1993 under two nitrogen treatments (i.e. N fertilized and no N fertilized). From Echarte [11].

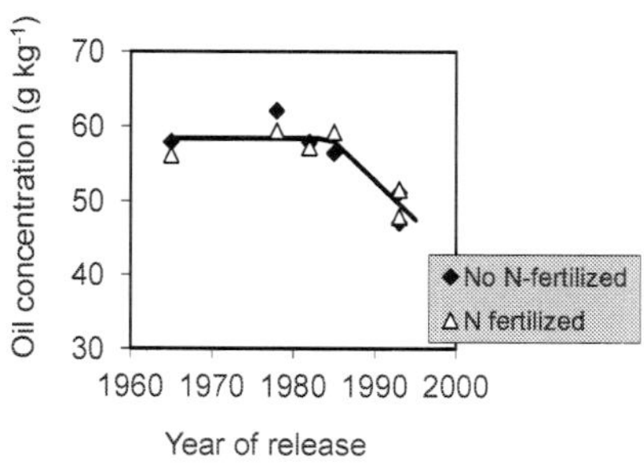


**Figure 9.** Oil concentration ($g\ kg^{-1}$) in grains of Argentinean maize hybrids released between 1965 and 1993 under two nitrogen treatments (i.e. N fertilized and no N-fertilized). From Echarte [11].

**Resource capture and resource use efficiency**

Greater grain yields of newer maize hybrids might have resulted in a concomitant increase in resource capture and/or resource use efficiency.

In non-limiting environments, grain yield can be expressed as the result of intercepted radiation, radiation use efficiency for shoot biomass production and harvest index [86]. Intercepted radiation did not consistently change in Argentinean maize hybrids released between 1965 and 1993 [11], in accordance with the lack of a consistent trend with the year of the hybrid release for shoot biomass. On the contrary, another study [40] reported accumulated intercepted radiation increments for Argentinean maize hybrids released between 1965 and 1997; which were attributed mainly to greater interception during the grain filling period. Contrasting results between works could be related to period under study and/or interaction between genotype and environment [87]. Nevertheless, grain yield increments were attributed to a large extent to greater radiation use efficiency for grain yield in both studies [11; 40]. The improved radiation use efficiency was not related to an improved light distribution within the canopy, as a lower extinction coefficient was not evident with the year of the hybrid release [40]. These results are in contrast to the more upright leaf habit with the year of the hybrid release reported for US hybrids [31]. A greater radiation use efficiency was also the main mechanism contributing to explain the greater shoot biomass of newer Canadian maize hybrids [88]. A smaller decline in maximum leaf photosynthetic rate from silking to maturity was the underlying process contributing to explain the greater radiation use efficiency in newer Canadian maize hybrids [16;17; 89]. Maximum leaf photosynthetic rates at silking, however, were similar among hybrids released in different decades [17].

In water and/or nitrogen limited environments, greater grain yields associated with resource capture increments might have exposed current maize hybrids to more frequent nutrient or water stresses. Nevertheless, as it was previously discussed, newer genotypes are more tolerant to stresses than older hybrids. Grain yields of newer maize hybrids were greater than those of older hybrids across **N levels** [31; 53; 90; 91]. Nitrogen use efficiency (the ratio of grain production to soil available N) can be expressed as the result of nitrogen recovery efficiency (NRE, the ratio of N uptake to soil available N) and nitrogen internal efficiency (NIE, the ratio of grain yield to whole plant N uptake at physiological maturity). Nitrogen use efficiency increased with the year of the hybrid release in Argentina during the 1965 – 2010 period [91; 92]. Nitrogen internal efficiency rather than greater N uptake largely explained the greater N-use efficiency of newer maize hybrids than older hybrids [92; 93]. These results are in agreement with findings in Canada and US [94]. The greater N-use efficiency in a newer than in an older Canadian maize hybrid was associated with a lower rate of decline of leaf photosynthesis towards physiological maturity, under both high and low N availability [17].

Water stress is one of the main limitations to crop grain yield worldwide; and it may reduce maize grain yield by 12-15% in temperate regions [95, 96]. Grain yield of newer maize hybrids was greater than that of older hybrids across **water regimes** during the grain filling period [97]. Preliminary results of our group indicate that grain yield improvements in

Argentinean maize hybrids released between 1980 and 2004 has been associated with increased water use efficiency for grain production and not with water uptake, which has remained relatively stable [98]. This is in contrast with previous reports suggesting that water capture increased with the year of the hybrid release in US hybrids [99]. However, the consistently increased total shoot biomass with the year of the hybrid release in US hybrids and not in Argentinean maize hybrids may contribute to explain discrepancies between works. Although seasonal water uptake was similar among Argentinean maize hybrids released in different decades, soil water uptake during the critical period for kernel set was greater in newer than in older maize hybrids when soil available water was low [100]. In agreement, a modern Canadian hybrid was able to maintain higher leaf photosynthesis and transpiration during short periods of low water availability at silking than an older hybrid in a greenhouse study [55]. Water use efficiency for grain production was consistently higher in a newer than in an older Argentinean maize hybrid, and differences were greater at low water availability [101].

## 2. Conclusions

Greater grain yield of newer Argentinean maize hybrids was mainly related to an increased harvest index; whereas shoot biomass did not consistently increased with the year of the hybrid release. Kernel number was the main yield numerical component contributing to explain grain yield increments. Processes influencing kernel number determination in hybrids released in different decades were analyzed using as a framework the relationship between kernel number per plant (KNP) and plant growth rate during the critical period for kernel set (PGRs); and it was evident that features of the relationship were changed through the years. As such, threshold PGRs for kernel set was lower and maximum kernel number per plant was higher in newer than in older hybrids. The lower threshold PGRs for kernel set contributed to explain the greater tolerance of newer hybrids to high plant densities, and it probably contributed to a greater tolerance to other stresses like low water availability or low soil N. The lower threshold PGRs for kernel set was associated with a greater assimilate partitioning to the ear at low resource availability per plant; which was probably related to a lower apical dominance in newer than in older maize hybrids. The higher maximum kernel number per plant at high resource availability was associated with morphogenetic changes leading to a greater potential kernel number per ear; whereas prolificacy was not consistently improved. This response of kernel number to an increased resource availability contributed to explain the greater grain yield of newer hybrids at low plant densities. As such, harvest index of newer maize hybrids was not only greater but it was also more stable at different resource availability than that of older maize hybrids.

Kernel weight did not show a clear trend with the year of the hybrid release; but it was evident that kernel weight of newer hybrids was more susceptible to stresses during the grain filling period than that of the older hybrids. Kernel weight response to resource availability during the grain filling period was analyzed in terms of the source-sink ratio. The sink or the ear demand for assimilates during the grain filling period was greatly increased in newer maize hybrids; as a result of either a greater kernel number and/or a greater kernel growth rate. Kernel growth rate has been shown to be closely associated with the potential kernel weight. However, the ear demand was increased to a greater extent than the source (i.e. plant growth during the grain filling period), and stresses during the grain filling period reduced kernel weight of newer hybrids more than that of the older hybrids. Thus, future breeding efforts for yield improvement would need to focus also on an increase in source capacity during the grain filling period. Kernel chemical quality was also modified with the year of the hybrid release; as such, protein concentration in kernels was lower in newer hybrids at moderate soil N availability. This change was attributed to both, the genotypes used in the selection programs as well as the increased carbon fluxes to the kernels without proportional increments in the flux of nitrogen. When N luxury consumption occurred, protein concentration in kernels was similar among hybrids released in different decades. Oil concentration also decreased in newer hybrids released in 1993.

Resource use efficiency increments, rather than greater resource capture, concomitantly increased with grain yield of Argentinean maize hybrids released in different decades. In non-limiting environments, radiation use efficiency for grain production (i.e. grain yield per unit of intercepted radiation) was a consistently increased mechanism contributing to explain the greater grain yield of newer maize hybrids. This was associated mainly with the greater partitioning of assimilates to the ear and/or a greater potential kernel number per ear that allowed for an increased harvest index. An improved light distribution within the canopy was not evident. In soil N-limited environments, greater yield of newer maize hybrids were associated with greater nitrogen use efficiency for grain production; which was largely explained by a greater nitrogen internal efficiency (i.e. the ratio of grain yield to whole plant N uptake at physiological maturity). Similarly, in water limited environments, water use efficiency for grain production was greater in a newer than in an older maize hybrid. The lower thresholds PGRs for kernel set in newer compared with older maize hybrids might have resulted in a lower frequency of barren plants or with low number of kernels and thus in a greater kernel number at low N or low water availability. Resource capture was not consistently increased with the year of the hybrid release indicating that stressful conditions are not more frequent in current maize hybrids than before.

## Author details

Laura Echarte, Lujan Nagore, Javier Di Matteo and Mariana Robles
*Research Council of Argentina (CONICET), Argentina*

Laura Echarte, Matías Cambareri and Aída Della Maggiora
*INTA Balcarce - Facultad de Ciencias Agrarias, Universidad Nacional de Mar del Plata. CC 276, 7620 Balcarce, Argentina*

## Acknowledgement

This work was supported by the Research Council of Argentina (CONICET), Agencia Nacional de Promoción Científica y Tecnológica (ANPCyT) and INTA.

## 3. References

[1] Duvick DN. The Contribution of Breeding to Yield Advances in maize (Zea mays L.). Advances in Agronomy 2005; 86 83-145.

[2] Duvick DN., Smith JSC., Cooper M. Long-term selection in a commercial hybrid maize breeding program. In: Janick J. (ed.) Plant Breeding Reviews 2004; 24 109-151.

[3] Duvick DN. What is yield? In: Edmeades GO., Banziger M., Mickelson HR., Peña-Valdivia CB. (eds.) Developing Drought- and Low N-Tolerant Maize. Proceedings of a Symposium. March 25–29, 1996 CIMMYT, El Batan, Mexico; 1997. p332-335.

[4] Cunha Fernandes JS., Franzon JF. Thirty years of genetic progress in maize (Zea mays L.) in a tropical environment. Maydica 1997; 42 21–27.

[5] Eyhérabide GH., Damilano AL., Colazo JC. Genetic gain for grain yield of maize in Argentina. Maydica 1994; 39 207–211.

[6] Eyhérabide GH., Damilano AL. Comparison of genetic gain for grain yield of maize between the 1980s and 1990s in Argentina. Maydica 2001; 46 277–281.

[7] Lee EA., Tollenaar M. Physiological basis of successful breeding strategies for maize grain yield. Crop Science 2007; 47(S3) S202–S215.

[8] Tollenaar M., Lee EA. Strategies for Enhancing Grain Yield in Maize. In: Janick J. (ed) Plant Breeding Reviews 2011; 34 37-83.

[9] Echarte L., Andrade FH. Harvest index stability of Argentinean maize hybrids released between 1965 and 1993. Field Crops Research 2003; 82 1-12.

[10] Echarte L., Andrade FH., Vega CRC., Tollenaar M. Kernel number determination in Argentinean maize hybrids released between 1965 and 1993. Crop Science 2004; 44 1654-1661.

[11] Echarte L. Yield determination in Argentinean maize hybrids released in different decades. PhD. Thesis. National University of Mar del Plata, Argentina; 2003. In Spanish.

[12] Tollenaar M., Lee EA. Dissection of physiological processes underlying grain yield in maize by examining genetic improvement and heterosis. Maydica 2006; 51 399–408.

[13] Crosbie TM. Changes in physiological traits associated with long-term breeding efforts to improve grain yield of maize. In: Loden HD., Wilkinson D. (eds.) Proceedings of the Annual Corn Sorghum Industry Research Conference. 37th. Chicago, IL. December 5–9 1982. American Seed Trade Association Washington, DC; 1982. p206–233.

[14] Tollenaar M., Muldoon JF., Daynard TB. Differences in rates of leaf appearance among maize hybrids and phases of development. Plant Science 1984; 642 759-763.

[15] Tollenaar M., Dwyer LM., Stewart DW. Physiological parameters associated with differences in kernel set among maize hybrids. In: Westgate MA., Boote KJ. (eds.) Physiology and modeling kernel set in maize. CSSA Special. Publication 51. CSSA/ASA/SSSA, Madison, WI. 2000; p115-130.

[16] Ying J., Lee EA., Tollenaar M. Response of maize leaf photosynthesis to low temperature during the grain-filling period. Field Crops Research 2000; 68 87–96.

[17] Echarte L., Rothstein S., Tollenaar M. The Response of Leaf Photosynthesis and Dry Matter Accumulation to N Supply in an Older and a Newer Maize Hybrid. Crop Science 2008; 48 656-665.

[18] Tollenaar M., Dwyer LM., Stewart DW. Ear and kernel formation in maize hybrids representing three decades of grain yield improvement in Ontario. Crop Science 1992; 32 432–438.

[19] Echarte L., Luque S., Andrade FH., Sadras VO, Cirilo AG., Otegui ME., Vega CRC. Response of maize kernel number to plant density in Argentinean Hybrids released between 1965 and 1993. Field Crops Research 2000; 68 1-8.

[20] Andrade FH., Cirilo A., Echarte L. Kernel number determination in maize. In: Otegui, ME., Slafer, G. (eds.) Physiological basis for maize improvement; 2000. p59-70.

[21] Tollenaar M. Sink-source relationships during reproductive development in maize. A review. Maydica 1997; 22 49-75.

[22] Edmeades GO., Daynard TB. The relationship between final yield and photosynthesis at flowering in individual maize plants. Canadian Journal of Plant Science 1979; 59 585-601.

[23] Tollenaar M., Daynard TB. Relationship between assimilate source and reproductive sink in maize grown in a short season environment. Agronomy Journal 1978; 70 219-223.

[24] Fischer KS, Palmer AFE. Tropical maize. In: Goldsworthy PR., Fischer NM. (eds.) The physiology of tropical field crops; 1984. p213-248.

[25] Kiniry JR., Ritchie JT. Shade-sensitive interval of kernel number of maize. Agronomy Journal 1985; 77 711-715

[26] Aluko GK., Fischer KS. The effects of changes of assimilate supply around flowering on grain sink and yield of maize (Zea mays) cultivars of tropical and temperate adaptation. Australian Journal of Agricultural Research 1988; 39 153-161

[27] Otegui ME., Bonhomme R. Grain yield components in maize. I. Ear growth and kernel set. Field Crops Research 1998; 56 247-256.

[28] Andrade FH., Vega CRC, Uhart SA., Cirilo AG., Cantarero M., Valentinuz OR. Kernel number determination in maize. Crop Science 1999; 39 453-459.

[29] Vega CRC., Andrade FH., Sadras VO., Uhart SA., Valentinuz OR. Seed number as a function of growth. A comparative study in soybean, sunflower and maize. Crop Science 2001; 41 748-754.

[30] Doebley J., Stec A., Hubbard L. The evolution of apical dominance in maize. Nature 1997; 386 485-488.

[31] Duvick DN. Genetic contributions to yield gains of U.S. hybrid maize, 1930 to 1980. In: Fehr WR. (ed.) Genetic contributions to yield gains of five major crop plants. CSSA Special Publication 7. Madison, WI; 1981. p15-47.

[32] Russell WA. Agronomic performance of maize cultivars representing different eras of maize breeding. Maydica 1984; 29 375-390.

[33] Troyer AF., Roosenbrook RW. Utility of higher plant densities for corn performance testing. Crop Science 1983; 23 863-867.

[34] Troyer AF. Breeding widely adapted, popular maize hybrids. Euphytica 1996; 92 163-174.

[35] Reeder LR. Breeding for yield stability in a commercial program in the USA. In: Edmeades, GO., Bänziger, B., Mickelson, HR., Pena-Valdivia, CB. (eds.). Developing

drought and low N tolerant maize. Proceedings of a Symposium. March 25–29, 1996. CIMMYT, El Batan, Mexico;1997. p387-391.

[36] Tollenaar M., Wu J. Yield improvement in temperate maize is attributable to greater stress tolerance. Crop Science 1999; 39 1597-1604.

[37] Tollenaar M., Lee EA. Yield potential, yield stability and stress tolerance in maize. Field Crop Research 2002; 75 161-169.

[38] Fasoula VA., Fasoula DA. 2002. Principles underlying genetic improvement for high and stable crop yield potential. Field Crop Research 2002; 75 191-209.

[39] Araus JL., Serret MD., Edmeades GO. Phenotyping maize for adaptation to drought. Frontiers in physiology 2012; 3 1-20.

[40] Luque SF., Cirilo AG., Otegui ME. Genetic gains in grain yield and related physiological attributes in Argentine maize hybrids. Field Crops Research 2006; 95 383–397.

[41] Vega CRC., Andrade FH., Sadras VO. Reproductive partitioning and seed set efficiency in soybean, sunflower and maize. Field Crops Research 2001; 72 163-175.

[42] Bolaños J., Edmeades GO. Eight cycles of selection for drought tolerance in lowland tropical maize. II. Responses in reproductive behaviour. Field Crops Research 1993; 31(3-4) 253-268.

[43] Edmeades GO., Bolaños J., Hernández M., Bello S. Causes for silk delay in a lowland tropical maize population. Crop Science 1993; 33 1029-1035.

[44] Cárcova J., Urribelarrea M., Borrás L., Otegui ME., Westgate ME. Synchronous pollination within and between ears improves kernel set in maize. Crop Science 2000; 40 1056-1061.

[45] Cárcova J., Otegui ME. Ear temperature and pollination timing effects on maize kernel set. Crop Science 2001; 41 1809-1815.

[46] Tollenaar M. Physiological basis of genetic improvement of maize hybrids in Ontario from 1959 to 1988. Crop Science 1991; 31 119-124.

[47] Tetio-Kagho F., Gardner FP. Responses of maize to plant population density II. Reproductive development, yield and yield adjustment. Agronomy Journal 1988; 80 935-940.

[48] Hashemi-Dezfouli A., Herbert SJ. Intensifying plant density response of corn with artificial shade. Agronomy Journal 1992; 84 547-551.

[49] Daynard TB., Muldoon JF. 1983. Plant - to - plant variability of maize plants grown at different densities. Canadian Journal of Plant Science 1983; 63 45-59

[50] Di Matteo JA., Cerrudo AA, Robles M., De Santa Eduviges JM., Rizzalli R., Di Benedetto, A., Andrade FH. Ecofisiología del rendimiento en híbridos de maíz (Zea mays L.) liberados en las últimas 2 décadas. IX Congreso Nacional de Maíz, Rosario, Argentina 2010.

[51] Tollenaar M., Aguilera A., Nissanka SP. Grain yield is reduced more by weed interference in an old than in a new maize hybrid. Agronomy Journal 1997; 89 239–246.

[52] Dwyer LM., Tollenaar M. Genetic improvement in photosynthetic response of hybrid maize cultivars, 1959 to 1988. Canadian Journal of Plant Science 1988; 69 81–91.

[53] Castleberry RM., Crum CW., Krull F. Genetic yield improvement of U.S. maize cultivars under varying fertility and climatic environments. Crop Science 1984; 24 33–36.

[54] Rajcan I., Tollenaar M. Source: sink ratio and leaf senescence in maize: I. dry matter accumulation and partitioning during grain filling. Field Crops Research 1999; 60 245-253.

[55] Nissanka SP., Dixon MA. Tollenaar M. Canopy gas exchange response to moisture stress in old and new maize hybrid. Crop Science 1997; 37 172-181.

[56] Andrade FH., Echarte L., Rizzalli R., Della Maggiora AI., Casanovas M. Kernel number prediction under nitrogen or water stress. Crop Science 2002; 42 1173-1179.

[57] Echarte L., Tollenaar M. Kernel set in maize hybrids and inbred lines exposed to stress. Crop Science 2002; 46 870-878.

[58] Duncan WG., Hatfield AL., Ragland JL. The growth and yield of corn. II. Daily growth of corn kernels. Agronomy Journal 1965; 57 221-223.

[59] Johnson DR., Tanner JW. Calculation of the rate and duration of grain filling in corn (Zea mays L.). Crop Science 1972; 12 485-486.

[60] Echarte L., Andrade FH., Sadras VO., Abbate P. Kernel weight and post flowering source manipulation in Argentinean maize hybrids released in different decades. Field Crops Research 2006; 96 307-312.

[61] Reddy VH, Daynard TB. Endosperm characteristics associated with rate of grain filling and kernel size in corn. Maydica 1983; 28 339-355.

[62] Jones RJ., Roessler J., Ouattar J. Thermal environment during endosperm cell division in maize: Effects on number of endosperm cells and starch granules. Crop Science 1985; 25 830-834.

[63] Jones RJ., Schreibe BM., Roessler J. Kernel sink capacity in maize: Genotypic and maternal regulation. Crop Science 1996; 36 301-306.

[64] Edmeades GO., Lafitte HR. Defoliation and plant density effects on maize selected for reduced plant height. Agronomy Journal 1993; 85 850-857.

[65] Uhart SA., Andrade FH. Nitrogen and carbon accumulation and remobilization during grain filling in maize under different source/sink ratios. Crop Science 1995; 35 183-190.

[66] Maddonni GA., Otegui ME., Bonhomme R. Grain yield components in maize II. Postsilking growth and kernel weight. Field Crops Research 1998; 56 257-264.

[67] Borrás L., Otegui ME. Maize kernel weight response to postflowering source-sink ratio. Crop Science 2001; 49 1816-1822.

[68] Borrás L., Slafer GA, Otegui ME. Seed dry weight response to source–sink manipulations in wheat, maize and soybean: a quantitative reappraisal. Field Crops Research 2004; 86 131-146.

[69] Uhart SA., Andrade FH. Source sink relationship in maize grown in a cool temperature area. Agronomie 1991; 11 863-875.

[70] Cirilo AG., Andrade FH. Sowing date and kernel weight in maize. Crop Science 1996; 36 325-331.

[71] Andrade FH., Ferreiro MA. Reproductive growth of maize, sunflower and soybean at different source levels during grain filling. Field Crops Research 1996; 48 155-165.

[72] Di Matteo J., Robles M., Cerrudo A., Rizzalli R., Echarte L., Andrade FH. Ear demand in Argentinean maize hybrids as affected by plant density and year of release. In:

Proceeding of the ASA, CSSA, and SSSA International Annual Metting. Cincinatti, Ohio, EEUU; 2012.

[73] Uhart SA., Echeverrría HE. Diagnóstico de la fertilización. In: Andrade, FH., Sadras, VO. (eds.) Bases para el manejo del maíz, el girasol y la soja. INTA-National University of Mar del Plata, Balcarce; 2000. p235-268.

[74] Dudley JW., Lambert RJ., Alexander DE. In: Dudley JW. (ed.) Seventy generations of selection for oil and protein concentration in the maize kernel. Crop Science Society of America; 1974. p181-212.

[75] Kamprath EJ., Moll RH., Rodriguez N. Effects of nitrogen fertilization and recurrent selection on performance of hybrid populations of corn. Agronomy Journal 1982; 74 955-958.

[76] Duvick DN., Cassman KG. Post-green revolution trends in yield potential of temperate maize in the north-central United States. Crop Science 1999; 39 1622-1630.

[77] Borrás L., Curá JA., Otegui ME. Maize kernel composition and post flowering source-sink ratio. Crop Science 2002; 42 781-790.

[78] Slafer G., Andrade FH., Feingold S. Genetic improvement of bread wheat (Triticum aestivum L.) in Argentina: relationships between nitrogen and dry matter. Euphytica 1990; 50 63-71.

[79] Calderini DF, Torres-León S., Slafer G. Consequences of wheat breeding on nitrogen and phosphorus yield, grain nitrogen and phosphorus concentration and associated traits. Annals of Botany 1995; 76 315-322.

[80] Vyn TJ., Tollenaar M. Changes in chemical and physical quality parameters of maize grain during three decades of yield improvement. Field Crops Research 1998; 59 135-140.

[81] Streeter JG., Barta AL. Nitrogen and minerals. In: Tesar MB.(ed.) Physiological basis of crop growth and development. ASA, CSSA Madison, Wisconsin; 1984. p175-200.

[82] Uhart SA. Deficiencias de nitrógeno en maíz: efectos sobre el crecimiento, desarrollo y determinación del rendimiento. PhD Thesis. National University of Mar del Plata, Balcarce, Buenos Aires, Argentina; 1995.

[83] Ingle J., Beitz D., Hageman RH. Changes in Composition during Development and Maturation of Maize Seeds. Plant Physiology 1965; 40 835-839.

[84] Perry TW. Corn as a livestock feed. In: Sprague GF., Dudley JW.(eds). Corn and Corn mprovement. American Society of Agronomy, Madison, WI; 1988. p941–963

[85] Lambert RJ., Alexander DE., Mollring EL., Wiggens B. Selection for increased oil concentration in maize kernels and associated changes in several kernel traits. Maydica 1997; 42 39-43.

[86] Gardner BR., Pearce RB., Mitchel RL. In: Gardner BR., Pearce RB., Mitchel RL. (eds.) Physiology of crop plants. Iowa State University Press; 1985. pp327.

[87] Turner N. Further progress in crop water relations. Advances in Agronomy 1997; 58 293-338.

[88] Tollenaar M., Aguilera A. Radiation use efficiency of an old and a new maize hybrid. Agronomy Journal 1994; 84 536-541.

[89] Ying J., Lee EA., Tollenaar M. Response of maize leaf photosynthesis during the grain-filling period of maize to duration of cold exposure, acclimation, and incident PPFD. Crop Science 2002; 42 1164–1172.

[90] Sangoi L., Ender M., Guidolin AF., Almeida ML., Konflanz VA. Nitrogen fertilization impact on agronomic traits of Maize hybrids released at different decades. Pesquisa Agropecuaria Brasileira 2001; 36 757-764.

[91] Ding L., Wang KJ., Jiang GM., Biswas DK., Xu H., Li LF., Li YH. Effects of nitrogen deficiency on photosynthetic traits of maize hybrids released in different years. Annals of Botany 2005; 96 925–930.

[92] Lahitte M., Uhart SA., Andrade FH. Eficiencia de uso de nitrógeno en híbridos de maíz liberados en distintas épocas en Argentina. VI Congreso Nacional de maíz. Pergamino, Buenos Aires, Argentina 1997; 3 129-136.

[93] Robles M., Cerrudo AA., Di Matteo JA., Rizzalli R., Andrade FH. Nitrogen use efficiency of maize hybrids released in different decades. ASA, CSSA and SSSA, International Annual Meetings. San Antonio, Texas, USA; 2011.

[94] Ciampitti IA., Vyn TJ. Physiological perspectives of changes over time in maize yield dependency on nitrogen uptake and associated nitrogen efficiencies: A review. Field Crops Research 2012; 133 48-67

[95] Edmeades GO., Cooper M., Lafitte R., Zinselmeier C., Ribaut JM., Habben JE., Löffler C., Bänziger M. Abiotic Stresses and Staple Crops. In: Nösberger J., Geiger HH., Struik PC. (eds.) Crop Science Progress and Prospects; 2001; p 137-154.

[96] Eyherabide GH., Guevara E., Totis de Zeljkovich L. Efecto del estrés hídrico sobre el rendimiento de maíz en la Argentina. In: Edmeades GO., Banziger M., Mickelson HR., Peña–Valdivia CB. (eds.) Developing Drought- and Low N-Tolerant Maize. Proceedings of a Symposium. CIMMYT, México, March 25-29 1996; p24-28.

[97] Campos H., Cooper M., Habben JE., Edmeades GO., Schussler JR. Improving drought tolerance in maize: A view from industry. Field Crops Research 2004; 90 19–34.

[98] Nagore ML., Echarte L., Della Maggiora AI., Andrade FH. 2012. Seasonal crop evapotranspiration in modern and older maize hybrids. ASA, CSSA, and SSSA International Annual Metting. Cincinatti, Ohio, EEUU; 2012.

[99] Hammer GL., Dong Z., McLean G., Doherty A., Messina C., Schussler J., Zinselmeier C., Paszkiewicz S., Cooper M. Can Changes in Canopy and/or Root System Architecture Explain Historical Maize Yield Trends in the U.S. Corn Belt. Crop Science 2009; 49(1) 299-312.

[100] Nagore ML., Echarte L., Della Maggiora AI., Andrade FH. Rendimiento y evapotranspiración en híbridos de maíz de diferentes épocas. Reunion Argentina de Agrometeorologia 2012b. October 17-19, 2012, Malargue, Mendoza; 2012.

[101] Nagore ML., Echarte L., Della Maggiora AI., Andrade FH. Respuesta de la fotosíntesis al estrés hídrico en híbridos de maíz. XXVIII Reunión Argentina de Fisiología Vegetal. La Plata September 26-29, 2010. Buenos Aires, Argentina; 2010.

Chapter 3

# Ecophysiology of Wild Plants and Conservation Perspectives in the State of Qatar

Bassam T. Yasseen and Roda F. Al-Thani

Additional information is available at the end of the chapter

http://dx.doi.org/10.5772/55305

## 1. Introduction

State of Qatar is a peninsula extended from Arabia desert as outcrop in the western Arabian Gulf, located in an area of the world which is warm and humid; its land is considered as arid or semi-arid and highly saline. The common type of landscape is rocky desert, depressions and salt marshes, and in general it is flat to undulating. Such environment could have great impact on the biodiversity of flora and fauna, since limited number of wild plants and animals has been recorded in this area. The vegetation of Qatar is comprised of herbaceous plants, dwarf shrubs and a few tree species. Woody plants, succulents, and perennial grasses and sedges withstand successfully the severe drought of the summer. Soil in Qatar is generally shallow sandy calcareous, overlying rocky bed rock. The available nutrition for native plants is poor with salty soil; they are adapted and tolerate different physical and chemical factors. These adaptations also affect the type, abundance, and occurrence of microorganisms. However, this country is endowed with natural resources especially gas and oil; their revenues have been used to support all aspects of social life including the expansion of urban and industrial sectors as well as supporting the scientific research and its international obligations. In fact, Qatar has engaged with numerous international activities such as humanitarian aids and hosting international sport competitions and political meetings. The expansion in the industrial activities might put the ecosystem at real risk; some habitats are disappearing, and pollution of the environment could be another possible threat of such activities in the long run. Three challenges facing scientists and the decision makers in the State of Qatar: (1) restoration of the endangered habitats due to the industrial and urban activities, (2) phytoremediation of wastewater and soil which might be affected by gas and oil industry, and (3) establishment of wide scale institutes to deal with the modern techniques of gene technology to develop transgenic crop and native plants to cope with harsh and polluted environments. Therefore, this chapter of the book (Agricultural Chemistry, ISBN 980-953- 307-1002-1) is dedicated to address the efforts of restoration, phytoremediation and modern technologies in the State of Qatar. The following objectives have been suggested to discuss these efforts and to implement scientific solutions of the problems facing the environment in this region: (1) describe the environment and the plant wild life in the Gulf region in general and in particular in the State of Qatar, (2) understand the ecophysiology and mechanisms of adaptation of plants and microorganisms under drought and saline environments, (3) comprehend the impact of human and industrial activities on the natural habitats, and the restoration and phytoremediation efforts to maintain the environment, and (4) foresee the perspectives of genetic manipulation of crop and native plants, and using the possible gene bank of native plants to improve the phytoremediation processes.

## 2. The environment in the Gulf region

By definition of aridity, as a function of rainfall and temperature and the daily follow-up of meteorological data; the Arabian Gulf region is considered as arid or semi-arid. This region is absolutely among the warmest regions of the world; the temperatures in the summer season reach levels as high as 50 degrees Celsius or may be above. The general outlook of the mean rate of precipitation in this region showed clearly that the rain is scarce and not exceeds the rate of 152 mm per year [1]. Thus, such shortage in the rainfall in the Gulf region which is coincided with the high rate of evaporation in most of the days of the year with the presence of salt water, would make most of the lands as dry and very highly saline, reaching high values of electrical conductivities of the soil saturated extracts, ECe (about 200 milli-Siemens / cm) in salt marshes, Sabkhas, coastal line and even in some inland areas [2-4]. The main source of salinity at the coastal line is from the saline water of the Gulf, while the high values of ECe in the inland areas and Sabkhas are attributed to the intrusion of seawater into the underground waters [5].

The meteorological data in the State of Qatar that have been obtained from Doha (The capital of Qatar) airport for the last two decades confirmed the reality of aridity of the land (Table 1). Some reports have indicated that the rainfall in this region is irregular and variable in time and space and unpredictable. From the data reported by Batanouny [6], the average annual rainfall in Doha for 17 years was 78.1 mm, this figure fluctuates between 0.4 mm in 1962 and 302.8 mm in 1964. The temperature records, on the other hand, show that not a single month has mean temperature below 17.1 °C, and the mean minimum temperature does not drop below 12.7 °C. The absolute maximum temperature in January is 30.7 °C and the absolute minimum temperature is 3.8 °C, and there is never a danger of frost. Over the year, the maximum temperatures are recorded in July and August; however an exceptional highest record of air temperature (about 49 °C) at Doha was reported in June 1962.

The world meteorological organization (UN) and Hong Kong Observatory reports about the climatic records in the city of Doha have confirmed the above conclusions [7]. Also, considerable variations have been found in most parameters of the physical and chemical characteristics of soils including: soil texture, water content, pH of the soil extracts and ECe;

EC of the saturated soil extracts (Table 2). Such variations in those parameters are more obvious in ECe; which accompanied with the variation in $Na^+$ and $Cl^-$ ions content in those soils. In fact, the data have shown that the main elements that could cause salt stress in such soils are $Na^+$ and $Cl^-$ in addition to other cations like Ca2+, Mg2+ and $K^+$.

| Climatic elements | Annual values | Months of lowest values | Months of highest values |
|---|---|---|---|
| Average temp. (°C) | 27 | Jan: 17 | Jul-Aug: 35 |
| Average max. temp. (°C) | 31 | Jan: 20 | Jun-Jul: 40 |
| Average min. tem. (°C) | 22 | Jan: 13 | Jul-Aug: 30 |
| Absolute max. temp. (°C) | 47 | Jan – Feb: 32 | May-Aug: 47 |
| Absolute min. temp. (°C) | 1 | Jan: 1 | Jul: 23 |
| Average rainfall (mm) | 81 | Jun-Oct: 0 | Feb: 20 |
| Average morning RH (%) | 71 | Jun: 52 | Feb-Dec: 82 |
| Average afternoon RH (%) | 43 | Jun: 28 | Dec: 54 |
| Average wind speed (km/h) | 20 | Sep-Oct: 14 | Jun: 27 |

RH: Relative Humidity

**Table 1.** Climatic data of Doha city. Average annual values, average monthly lowest values, and average monthly highest values. Data of 17 years [5].

The above meteorological conditions and soil properties could be involved strongly to the features of the harsh environment in the region, and ultimately could have had major impact on the biodiversity and distribution of wild life in Qatar. Moreover, the shape, topography, geological structure and petrographic composition of the landscape are additional factors affecting the life of plants, as well as other living organisms [6]. Thus, from geographic point of view and climate conditions, Qatar is considered as hot subtropical desert with saline flats [5, 8].

| Parameters | Properties |
|---|---|
| Soil texture | Sandy - Sandy clay loam |
| Absolute water content (%) | 2.2 – 25.7 |
| Soil water content (% field capacity) | 7.9 – 65.9 |
| pH (soil extract) | 6.6 – 8.4 |
| ECe ($dSm^{-1}$) | 4.2 – 195.0 |
| $Na^+$ ($mg\ l^{-1}$) of soil extract | 186 – 726 |
| $K^+$ ($mg\ l^{-1}$) of soil extract | 60 – 108 |
| $Ca^{2+}$ ($mg\ l^{-1}$) of soil extract | 63 - 113 |
| $Mg^{2+}$ ($mg\ l^{-1}$) of soil extract | 34 – 82 |
| $Cl^-$ ($gl^{-1}$) of soil extract | 2.0 - 344 |

**Table 2.** The physical and chemical properties of soil samples collected from Doha area [9].

## 3. The wild plants and adaptations

Qatar is a small country with more than 4/5 of its boundary surrounded by a coastline (Figure 1), and the combinations of many environmental factors including drought and salinity, and high evapo-transpiration, irradiance, and temperatures do not allow many plants to grow and survive in this region. The flora of Qatar has been placed into one of five main habitat groups [10]: (1) xerophytes of rock and gravel deserts where the conditions are very dry, (2) halophytes of saline areas such as salt marshes, coastal sands, sabkhas and oolitic sands, (3) xerophytic species grow in natural silt and sand depressions where water retention is higher than those purely xerophytes, (4) species adapted at deep sand where water is available under the surface, and (5) species receiving artificial irrigation such as farms, gardens and sewage ponds. Also, the coastal line has been classified into five halophytic plant communities: (1) mangrove intertidal community, (2) low salt marsh coastal community, (3) high salt marsh coastal community, (4) sandy coastal community, and (5) sandy-rocky coastal community [5]. The coastline of the State of Qatar encompasses unique ecosystem of vegetated sand beaches alternating with bays of aquatic halophytes and mangroves, followed inland by members of some families like chenopodiaceae, which are considered as halophytes and/or xerophytes. These plants have various morphological and anatomical modifications, and physiological and biochemical characteristics that could have contributed to their adaptation to the harsh environment in this region [9].

The checklist of plants in the State of Qatar has been reviewed many times, some reports [11] listed 213 species, while others [12] gave a list of 260 species, then the monograph of Batanouny [6] listed 301 species in 207 genera and 55 families. Recently, Norton and his colleagues [10] have increased the list of wild plants to nearly 400 species of which about 270 species are likely to be truly native. Some other references can be found in the last publication to updating the checklist of the wild plants in the Gulf region and in the State of Qatar. Most of these plants have been recognized as either xerophytes or halophytes; well adapted to dry and/or saline environments.

### 3.1. Xerophytic plants

Many plants at the coastal line and throughout the Qatari land are characterized as xerophytes, include: *Cyperus conglomeratus, Oligomeris linifolia, Helianthemum lipii, Tetraena qatarense, Ochradenus baccatus* and many others. These plants have various mechanisms to cope with drought and water stressed soils

*(A) Drought - Escaping xerophytes*

These plants germinate, grow, and flower within a short period of time after fairly heavy rainfall. They produce seeds before the dry season, and therefore, they resist dry season during the seed stage [13, 14]. Examples of such plants: *Polycarpaea spicata* and *Senecio desfontanei*. Almost all ephemerals that germinate and grow in the desert after the summer rainfall are C4 plants, while plants that germinate after the autumn rainfall are C3 plants [13]. Most escaper plants survive the dry period in the desiccation – tolerant seed stage.

Some efforts have been made to develop crop plants that can complete life cycle in a very short period of time, to avoid the dry season. Genetic hybridization and selection of some barley cultivars are among these efforts [15-17].

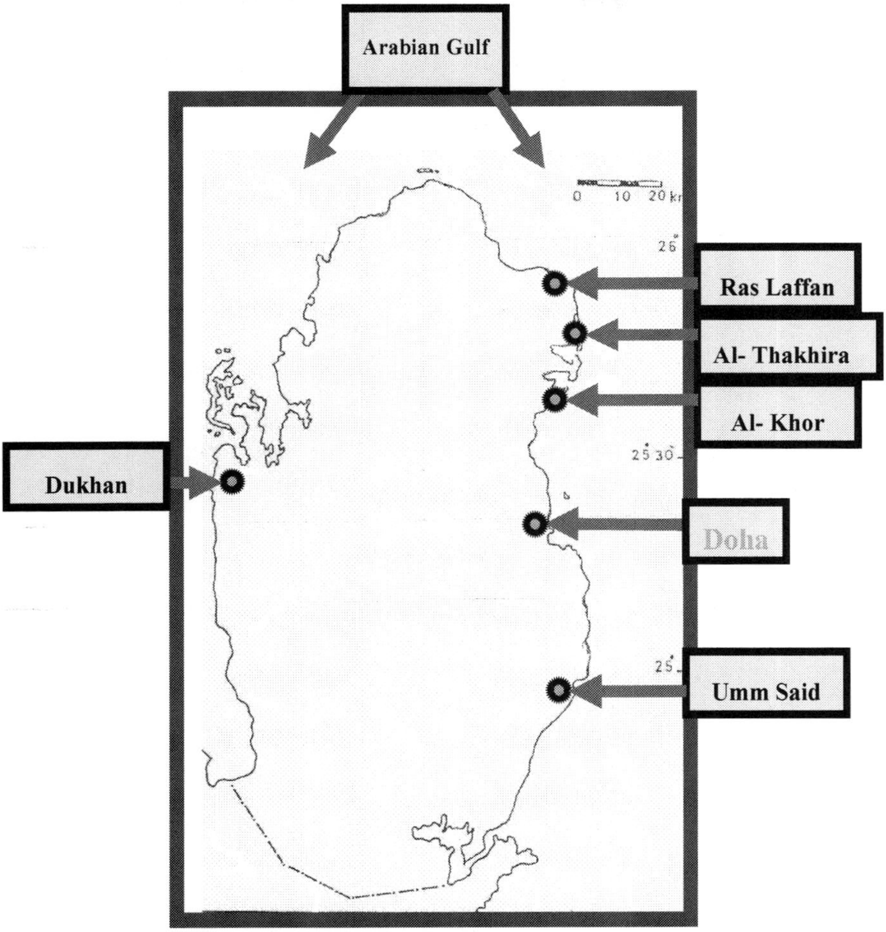


**Figure 1.** A map of Qatar showing the main cities and locations.

*(B) Drought avoidance*

These plants maintain favorable water content when exposed to external water stress. Drought avoidance is largely morphological-anatomical in nature. Two secondary mechanisms to avoid drought are considered here:

a. Water conservation: many plants can conserve water through the following adaptations:

1. Morphology and behavior of stomata: stomata are differing in plant species in their number and distribution on both surfaces of the leaf, size, shape and behavior. These features are species-specific characteristics, and vary with the adaptation to stress conditions. In *Cyperus conglomeratus*, stomata appear only on the abaxial surface of the leaves and these stomata are sunken [18]. Moreover, most plants living in the desert areas or saline soils close their stomata during periods of osmotic stress except CAM (Crassulacean Acid Metabolism) plants. However, many plants have special physiological and biochemical features to cope with the harsh environments [19-21]. Data from glycophytes have revealed considerable reduction in the stomatal conductance with salinity [22]. In this respect, it should be understood that stomatal closure may help maintain the water balance inside the plant. Thus, stomatal closure can be considered as an adaptation characteristic that plants have in various degrees under conditions of osmotic stress [22-24]. *Cyperus conglomeratus* exhibits C4 metabolism and this characteristic might allow the plant to maximize rates of photosynthesis by sustaining stomatal opening [25].
2. Increased cuticle thickness by increasing the surface lipid: the cuticle of plants under osmotic stress may become more thickened than those living under normal conditions (well irrigated or / and non-saline) [19, 22, 26]. The thickened cuticle is observed in many of these plants like *Cyperus conglomeratus* and *Tetraena qatarense* [18].
3. Decreased transpiring surface: the reduction in the leaf area is a common response to osmotic stress, and could be a main reason behind the reduction in the total transpiration rate in most plants [22, 27]. In fact, the reduction of leaf area at stressed growth conditions has been considered as a complex response and can be seen as an adaptation feature to reduce the water lost by transpiration process [28]. Some plant species in the Qatari flora like *Ochradenus baccatus* showed great reduction in the size of leaves to reduce the transpiring surface. In some other plants, rolling, folding, or shedding of leaves are possible methods of drought avoidance mechanisms in many desert plants [26]. *Aeluropus lagopoides* is another example of wild plants having some morphological modifications to cope with dry soils which include leaves linear; lanceolate with small blades ending in sharp rigid points [4]. These plants might have green stems to increase the photosynthetic efficiency under the severe environmental conditions [29].
4. Root adaptation: plant roots can contribute effectively to the drought avoidance mechanism by three ways: (a) restricting the root surface and decreasing its permeability to water, (b) quick development of roots to absorb rain water and disappear soon after soil dries, and (c) roots can reduce transpiration by high resistance to water [15, 30]. High Root / Shoot ratio, on the other hand, has been considered as a trait of a plant having drought avoidance mechanism [19, 31], *Cyperus conglomeratus* is a good example of such trait (Fig. 2) which helps withstand water stress by two ways: (1) less water is needed for the top, (2) exploring larger volume of soil.
5. Water storage: many plant species show succulence characteristics (Fig. 3); such plants could have water cells in stems, leaves and roots, which might confer avoidance mechanism against drought and water shortage. CAM plants are good example of how

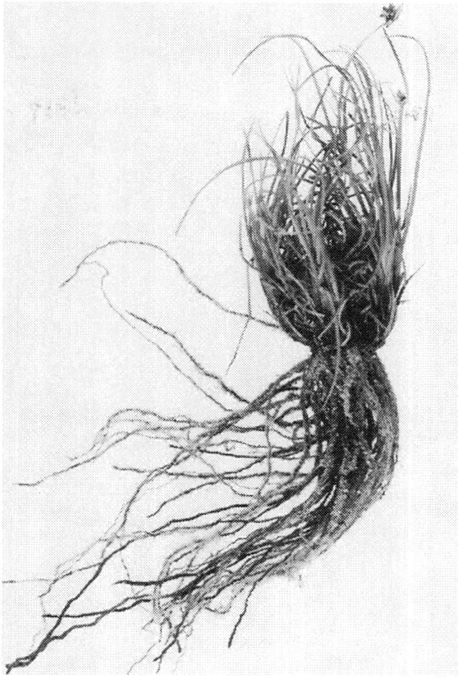

**Figure 2.** *Cyperus conglomeratus* has high root / top ratio.

plants can conserve water by closing stomata during the day to avoid water loss, and open them during the night for $CO_2$ fixation [32]. The flora of Qatar includes many succulent plants some of which exhibit CAM characteristics. The following species have been recognized as CAM plants: *Mesembryanthemum nodiflorum, Cyperus conglomeratus, Euphorbia granulate, Erodium laciniatum, Portulaca oleracea* and may be many others; these plants seem to have high water use efficiency associated with metabolic pathways that enable them to survive such environmental conditions [33, 34]. Some CAM criteria like diurnal oscillation of titratable acidity, malic acid content and succulence have been studied in three native plants of the flora of Qatar like *Salsola baryosma, Anabasis setifera,* and *Tetraena qatarense.* It was concluded that *A. setifera* is a constitutive CAM plant while *S. baryosma and T. qatarense* are inducible CAM plants [35]. The study of Mazen [36] on *Portulaca oleracea* ($C_4$ plant) may show two different behaviors concerning CAM characteristics, these are CAM-expressing and non CAM-expressing, and the induction of CAM-like characteristics in this plant was accompanied by increased activity and synthesis of phosphoenolpyruvate carboxylase (PEPC). *Mesembryanthemum nodiflorum,* the slender ice plant, is a succulent annual that is native to southern Africa, and introduced to many parts of the world. It favors saline and degraded soils of the agricultural regions, and it has the ability to switch from $C_3$ photosynthesis to CAM model during its growth [37]. *Oligomeris linifolia,* on the other hand, is not a CAM plant in spite of its succulence characteristic and its ability to store water in different plant parts (Fig. 4). This species has been considered as a desert species suspected as halophyte. Its life form was considered as Sub-Fruticose Chamaephytes, erect fruit shoot at base [38]. It grows in various habitats, including disturbed areas and saline soils, in deserts, plains, coastline, and other places. It is a fleshy annual plant, producing several erect, ribbed stems 35 to 45 centimeters in maximum height. Although these plants are considered as xerophytes, they might have also well adapted to saline environments, since they are living in soils of high salinity levels [4].

**Figure 3.** Branches of *Tetraena qatarense with* the fleshy leaves and woody stems.

b. Accelerated water absorption by water spenders: some plants can improve water uptake and content under drought by producing extensive roots or intensive fibrous root system to increase the active root surface area. *Helianthemum lipii* is a good example of a plant species having roots growing in the deeper soil layers toward the water table (Fig. 5). Also, some plants have high root / top ratio such as *Cyperus conglomeratus* (Fig. 2) which helps to cope with drought as mentioned before [15, 19, 26].

*(C) Drought tolerance*

Plants can tolerate drought (water stress) by two main mechanisms: (a) Dehydration avoidance. (b) Dehydration tolerance. Plants living in the State of Qatar are resistant to drought not only because of the morphological characteristics as avoidance mechanism, but also by osmotic adjustment [15, 39] through accumulation of substantial amounts of organic and inorganic solutes like proline, glycinebetaine, sugars and inorganic ions to cope with soil of severe water shortage [3, 40-42]. Plants such as *Tetraena qatarense* and *Ochradenus baccatus*, are good examples of xerophytes having dehydration avoidance mechanism by accumulating solutes to withstand severe water stress [4]. Dehydration tolerance mechanism, on the other hand, includes some methods like avoidance of starvation strain by stomatal opening at low $\Psi_w$, uncoupling of photosynthesis from transpiration, low respiration rate and other secondary mechanisms include: tolerance of starvation strain, avoidance of protein loss, and tolerance of protein loss [15]. Understanding these secondary mechanisms needs extensive and deep investigation in the plants covered in this review.

**Figure 4.** *Oligomeris linifolia* habit.

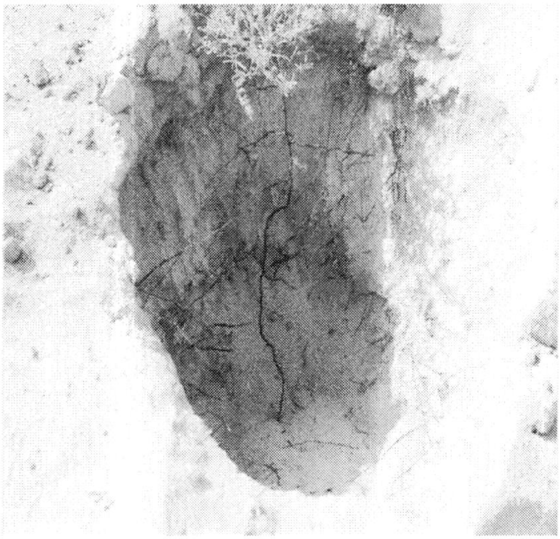

**Figure 5.** Some plants have roots growing in the deeper soil layers toward the water table, *Helianthemum lipii.*

## 3.2. Halophytic plants

Halophyte plants in Qatar are living near the coastal line, and in other inland areas like Subkhas, including some species of the genera *Anabasis, Arthrocnemum, Atriplex, Avicennia, Halocnemum, Halopeplis, Limonium, Salsola, Seidlitzia, Suaeda* and may be many others. These plants have two main mechanisms to cope with saline environments: (a) Avoidance mechanisms, (b) Tolerance mechanisms [9, 15, 26, 41, 43].

*(a) Avoidance mechanisms*

These mechanisms involve structural and physiological adaptations to minimize salt concentrations in the cells or physiological exclusion by root membranes. Plants can use one or more of the following secondary mechanisms to avoid salt stress at their environment: exclusion, extrusion and dilution.

i. Exclusion mechanism: plants having such mechanism may exclude ions at different locations along plant organs: (a) at the surface of the roots (b) between shoot system and root system, and (c) between leaves and petioles or sheaths. The best example of exclusion mechanism at the root surface was found in Date Palm trees (*Phoenix dactylefera*), which seems excluding $Na^+$ and $Cl^-$ from the root, by having a barrier at the root surface, since the uptake of $Na^+$ and $Cl^-$ was not proportional to the concentration of these ions in the external solution. The efficiency of exclusion increases with increasing salinity in the growth medium, and as a result, the accumulation of $Na^+$ and $Cl^-$ in the tissue was not much greater at high than at low salinity levels, and there were no visible symptoms of salt injury [15, 44]. Halophytes and may be many glycophytes exclude ions from leaves to stems or roots [20, 38]. Some reports [45] described the avoidance mechanism in the mangroves (*Avicennia marina*), by retaining low internal salinity in stems and leaves by means of salt excluding mechanisms in the roots. Leaves of this plant are grey and hairy at the lower surface where most secretion occurs, while the above surface is shiny green, glabrous with some salt glands. These salt glands play significant role in the internal balance and regulation of ions [9, 46]. In glycophytes, there are two examples from Mexican wheat plants, in which exclusion of $Cl^-$ ions to the sheaths can be considered as a regulatory mechanism in these plants to avoid $Cl^-$ accumulation in the leaves. Also, salt sensitive cultivars, like Yecora, failed to exclude $Na^+$ to the root systems as compared to salt resistant cultivars, like Cajeme, [22]. Moreover, salt exclusion may occur at the intracellular levels, and salt resistance can be attributed to the maintenance of ions homeostasis in the cytoplasm [47].

ii. Salt extrusion: it is an active process to excrete extra salts from the epidermal hairs or salt glands of leaves of some halophytes such as *Limonium axillare* and *Atriplex* spp. Some specialized structures called salt glands or salt bladders are found in leaves of those plants to regulate the extra salts inside the plant body [43]. Salt glands found in *Limonium axillare* are composed of (4-10) cells which are entirely covered by cuticle which seals the gland from the rest of the plant except for small gaps on the leaf mesophyll side, salts are deposited on the leaf surface through holes on these glands, and in mangroves (*Avicennia marina*), salts are excreted from the upper surface of leaves [46], and the excreted solutions exceed the NaCl concentration of seawater of about 10 times [38, 45, 48]. Salt Bladders are found in *Atriplex*, which are comprised of two cells: stalk cell and bladder cell [20, 49]. Salt glands are found also in many other halophytes like *Aeluropus lagopoides*, *Tamarix* spp. and may be in other plant species [9, 50, 51]. These salt glands could regulate the secretion of salts from leaves to keep their concentrations at low levels and ultimately to maintain ion homeostasis in the plant body.

iii. Salt dilution: some halophytes can dilute the accumulated ions in plant tissues to keep cytoplasmic salinity below toxic levels. Succulence is feature of the whole plant body or may be confined to stems and / or leaves. These plants are considered as salt includers. Some succulent halophytes living in Qatar include *Halopeplis* (Fig. 6), *Suaeda* (Fig. 7) and *Tetraena* (Fig. 3). Halophytes having dilution mechanism are characterized by: (1) thickening in leaves, (2) elongation of cells, (3) higher elasticity of cell wall, (4) smaller relative surface areas, (5) decrease in extensive growth, and (6) high water content per unit of surface areas [38, 52]. In fact, most halophytes accumulate inorganic ions that are found in abundance in the environment like $Na^+$ and $Cl^-$, and such accumulation is accompanied with a decline in $K^+$ content [41]. In addition to the ion accumulation and succulence phenomenon, shedding of old salt-saturated leaves is found in plants having dilution mechanism to avoid the damage caused by extra salt accumulation [53].

*(b) Tolerance mechanisms*

Osmoregulation or osmotic adjustment has been considered as a main secondary mechanism to tolerate salt stress. Osmotic adjustment can be defined as maintenance of positive water balance between soil environment and plant tissues, by lowering their plant

**Figure 6.** *Halopeplis perfoliata* habit.

**Figure 7.** *Suaeda aegyptiaca* shoots.

water and solute potentials (Table 3). In halophytes, this mechanism involves physiological and biochemical adaptations for maintaining protoplasmic viability as cells accumulate inorganic ions mainly $Na^+$, $Cl^-$, followed by $Ca^{2+}$, $K^+$ and $Mg^{2+}$, and other organic solutes to achieve osmotic adjustment [41, 46, 54]. However, these ions may be sequestered in vacuoles leaving relatively low ions in the cytoplasm. Some authors [55] reported in their review that crops including cereals tolerate NaCl by excluding $Na^+$ from the transpiration stream as well as sequestration of $Na^+$ and $Cl^-$ in the vacuoles of root and leaf cells, and promote other physiological processes like faster growth rates and longer duration by maintaining high concentrations in $K^+$ despite the osmotic stress of the salt outside the roots [28, 56]. Organic solutes like glycinebetaine, proline, sugars, polyols etc., may accumulate in the cytoplasm to achieve osmotic balance inside the cells, and therefore play a pivotal role in plant cytoplasmic osmotic adjustment. Thus, salt tolerance in these plants can be attributed mainly to the fact that these plants accumulate low-molecular-mass organic compounds (compatible solutes) [57-60].

| Salt stress of growth medium (NaCl, $molm^{-3}$) | Osmotic potential of growth medium ($\Psi_s$, MPa) | Roots | | Leaves | |
|---|---|---|---|---|---|
| | | Cajeme | Yecora | Cajeme | Yecora |
| 0 | -0.05 | -0.50 | -0.49 | -0.92 | -0.98 |
| 50 | -0.25 | -0.62 | -0.62 | -1.17 | -1.27 |
| 100 | -0.44 | -0.72 | -0.69 | -1.46 | -1.51 |
| 150 | -0.61 | -0.80 | -0.74 | -1.85 | -1.86 |

**Table 3.** Osmotic adjustment (MPa) in two Mexican wheat cultivars exposed to salt stress [22].

These solutes, as non-toxic cytoplasmic osmotica, that play major roles in the physiology and biochemistry of plant cells and have contributed in the process of osmotic adjustment, maintaining turgor and hydration of cellular microstructures, as sources of some carbon and nitrogen skeletons, and osmoprotectants [61, 62]. As osmoprotectants, these compounds tend to be excluded from the hydration sphere of proteins and stabilizing the folded protein structure [63-65], maintaining plasma membranes, protecting the transcriptional and translational machineries and intervening in the process of refolding of enzymes as molecular chaperones [66, 67]. Moreover, compatible compounds like proline and glycinebetaine can induce the expression of certain stress-responsive genes, including those for enzymes that scavenge reactive oxygen species [68].

## 4. Microorganisms associated with wild plants

Higher plants are normally colonized by microorganisms, which include bacteria, fungi, algae or protozoa. Microorganisms interact with plants because plants offer a wide diversity of habitats including the (a) phyllosphere (aerial plant part), (b) rhizosphere (zone of influence of the root system), and (c) endosphere (internal transport system). Interactions of epiphytes, rhizophytes or endophytes may be detrimental or beneficial for either the microorganism or the plant, and may be classified as neutralism, commensalism, synergism, mutualism, amensalism, competition or parasitism [69]. Symbiosis is a relationship that both microorganisms and plants get benefits, and nitrogen fixation and mycorhizae are good examples of such relationships. The plant provides carbon materials to support the growth of microorganisms, while the latter promote plant growth by enhancing minerals uptake e.g. nitrogen and phosphorus (70). A mini review by [71], about the symbiosis relationship between desert plants and Mycorrhizae, indicated that desert ecosystems were not different from other ecosystems in the presence of mycorrhizae. These mycorrhizae might affect nutrient acquisition such as P, N, Fe, Zn, K and others. They also increase plant adaptation to abiotic stresses and some other stresses. Commensalism is another example that various chemicals are secreted from various plant parts like roots and leaves to stimulate the growth of microorganisms. Some other microorganisms can cause some diseases to plants, and this happens when the natural defense systems of the plant are ineffective. In fact, plants may limit microbial penetration by having a thick cell wall and other structural barriers like the cuticle layer that restrict infection. Moreover, the defense system in the plant includes the secretion of gums and some chemicals to limit the invasion of microorganisms [72]. The following are some examples of microorganisms associated with various plants from the Qatari environment and the perspectives of using these organisms to solve the outstanding problems of health, economy and food security.

### 4.1. Mangrove

Qatar is home to *Avicennia marina*; it is known as the grey mangrove or white mangrove trees, communities of which form several forests around Qatar shores. These mangrove swamps are home to a wealth of life. The largest area of mangroves - and the oldest - can be found around Al Thakhira and Al Khor. Other mangrove areas in Qatar originate from fairly recent plantings by the government.

Decomposers play an important role in the cycling of material and the flow of energy through an ecosystem. In the mangrove ecosystem, bacteria and fungi break down dead organic matter, such as mangrove leaves. One teaspoon of mud from a mangrove forest is estimated to contain 10 billion bacteria. These bacteria break down the leaf litter and provide nutrients for the other organisms that live in the mangrove swamp [73, 74]. This forms the basis of the food chain in the mangrove swamp. The nutritional value of the leaves is increased by the work of decomposers. Mangrove ecosystems are an important natural resource that should be protected. The detritus generated by the mangroves is the base of an extensive food web that sustains numerous organisms of ecological and commercial importance. Furthermore, mangrove ecosystems provide indispensable shelter and nurturing sites for many marine organisms. The well-being of mangroves is dependent on the diverse, and largely unexplored, microbial and faunal activities that transform and recycle nutrients in the ecosystem. Conservation strategies for mangroves should consider the ecosystem as a biological entity [75]. Despite numerous studies on the biogeography, botany, zoology, ichthyology, environmental pollution, and economic impact of mangroves, little is known about the activities of microbes in mangrove waters and sediments. An effort must be made for further studies on microbial activities in mangrove ecosystems and their impact on the productivity of the ecosystem.

Various types of microorganisms are found around the mangrove habitats. For example, diverse cyanobacterial communities reside on leaf, root litter, live roots, and often form extensive mats on the surrounding sediments; many of these communities are capable of fixing atmospheric nitrogen. Many genera widespread in these habitats including *Oscillatoria, Lyngbya, Phormidium* and *Microcoleus*are, and as heterocystous genera, *Scytonema*is common in some areas [76]. The filamentous cyanobacterium *Microcoleus* sp.was isolated and inoculated on to young mangrove seedlings. Such cyanobacterial filaments colonize the roots of mangrove by gradual production of biofilm [77]. Also, bacteria may influence the mangrove ecosystem directly; they contribute inevitably in the recycling of nutrients [78]. Many potential bacteria were isolated from mangrove ecosystem such as nitrogen-fixers, phosphate solubilizers, photosynthetic anoxygenic sulfur bacteria, methanogenic and methane oxidizing bacteria, which are involved in efficient nutrient recycling [79]. Also bacteria and fungi that either produce antibiotics [80, 81] or resistant to antibiotics were isolated from mangrove ecosystem [82, 83].

## 4.2. Halophytes

Salt-tolerant microorganisms can grow in habitats containing high concentrations of salts. The natural environments for salt-tolerant microbes may be similar to those of the halophytic angiosperms. The chemical analysis of the materials excreted by the epidermal glands of halophytes revealed the presence of mineral elements and some organic compounds, such substances might be a source of nutrition to the microorganisms living on the plant body [84]. A study on the desert plants of Egypt [85] indicated that the fungal species inhabiting the surface of senescent leaves of the succulent halophyte *Zygophyllum album* L. appeared to be adapted to stressful conditions of their microhabitats, namely high convective heat, dry conditions and high salt content of their leachates. In the study of [84] on some halophyte species growing at the Qatar North East coast showed that the total bacterial count in the rhizosphere was higher than in the non-rhizosphere soil. Moreover, the bacterial counts in the soil supporting the plant species of the coastal zone were higher than those in the soil of the inland zones. Also, Gram- positive cocci predominated in the isolates from rhizosphere and non-rhizosphere soil, and the isolates with white colonies color predominated in the rhizosphere than from phyllosphere since low colonization of bacterial cells were found on the aerial parts that have high contents of mineral ions due to the activity of salt glands in most Halophytes. Also, the bacterial counts were higher on the green parts of the plant than on the senescent parts, since the latter might have accumulated high concentrations of mineral ions as a mechanism to exclude salt to the aged parts of the plants compared to the growing ones. Moreover, the phyllosphere of the green and senescing parts were characterized by the predominance of Gram-positive bacilli and by the low percentage of isolates producing colored colonies. The general conclusion that can drawn from various published studies that soil environment and phyllosphere of halophytes support the growth of bacteria which seemed to have various mechanisms to deal with the harsh environments like salt marshes and sabkhas. In fact, bacteria are the most abundant inhabitants of the phyllosphere and the most colonists of leaves. On the other hand, the bacterial flora of the above ground differs substantially from that at the subterranean plant surfaces. For example, the pigmented bacteria which are rarely found in the rhizosphere, and laminate leaf surfaces and solar radiation affect the ecology of the phyllosphere and promote bacteria to produce pigments as sun screen and will not damage the cell components. There are two general strategies for osmo-adaptation of prokaryotes like bacteria under osmotic stress conditions: (1) accumulation of inorganic ions, and (2) accumulation of low molecular weight organic molecules. In the first strategy, osmotic equilibrium is maintained which involves the selective influx of potassium ($K^+$) and chloride ($Cl^-$) into the cytoplasm. The extremely halophilic archaea of the family *Halobacteriaceae* and the bacterium *Salinibacter ruber* as well as the moderately halophilc bacteria of the order *Haloanaerobiales* accumulate enormous quantities of $K^+$ and $Cl^-$. The second strategy of osmo-adaptation, on the other hand, involves the accumulation of a limited range of low-molecular-weight organic solutes. These include many compounds which are called compatible solutes including amino acids like proline, glycinebetaine, simple sugars, polyols, and their derivatives. Some microorganisms, such as hyper/thermophiles, utilize a combination of both strategies by accumulating negatively charged compatible solutes [86] and potassium. Great deals of attention have been paid to proline and glycinebetaine as compatible solutes accumulated as a result of salt or water stress. Relatively few prokaryotes are capable of de novo synthesis of these compounds. The intracellular concentrations of these solutes can be regulated in accordance with the external salt concentration, provide microorganisms with a large degree of flexibility and the possibility to adapt to a wide range of salt concentrations. However, energetically the production of massive amounts of such solutes can be costly.

### 4.3. *Prosopis cineraria*

*Prosopis cineraria* is the only species that grow rarely in desert of Qatar. The genus *Prosopis* contains around 45 species of spiny trees and shrubs found in the subtropical regions of Americas, Africa, Western Asia, and South Asia. They often thrive in arid soil and are resistant to drought, on occasion developing extremely deep root systems. The microbiology analyses of *P. cineraria* trees in Qatar [87] showed that the bark was colonized by large number of bacteria as compared to the leaves. Gram- positive cocci and spore forming bacilli bacteria are characterized by thick cell wall that is comprised of peptidoglycan (amino acid polypeptide and a sugar), and the isolated bacilli genus were spore forming that can be survive in hot, dry conditions, and high irradiation with limited damage to the cell. These are the most dominant epiphytic form on both leaves and bark. The pigmented bacteria, red yellow, and orange, were isolated from leaves and bark which exposed to long duration of light, that pigments well keep the bacterial cells undamaged and resistant to irradiation. The observed high content of organic matter, soil nutrients, clay and moisture in the sub-canopy locations of *P. cineraria* trees significantly affect the richness of the below canopy sites [87]. The bacterial soil populations in the rhizosphere are higher than the non-rhizosphere sits and the lowest bacterial count occurred in the outer canopy soil. Moreover, the presence of plant litter, animal droppings together with the already existed soil chemical nutrients in the sub canopy positions of *P. cinerearea* trees possibly increase the enrichment of these patches.

One of the best documented spatial patterns of nutrient distributions in arid and semiarid ecosystems are the "islands of fertility" associated with shrubs and trees [88].

The components of rhizosphere system which include microorganisms, plant and soil interact with each other in a way so that the rhizosphere is distinguished from the bulk of the remaining soil. The activity of root microorganisms is affected by soil environmental factors or by environmental factors operating indirectly through the plant. Moreover, root microorganisms can affect the plant and plant nutrient uptake, directly by colonizing the root and modifying the soil environment around the root. Bacterial growth is stimulated by a vast range of organic materials released from plant roots, which include carbohydrates, vitamins, amino acids and enzymes. Organic acids and lipids reduce the pH of the rhizosphere and also have a role in the chelation of metals [89]. However, direct benefit to the plant growth is not easy to demonstrate but the activities essential to plant growth, including mineralization and nitrogen fixation by free–living bacteria are concentrated in the rhizosphere. Miscellaneous compounds including volatile substances can physiologically stimulate or inhibit organisms. There are many factors affecting the release of organic compounds include plant species and cultivars, age and stage of plant development, light intensity and temperatures, soil factors, plant nutrients, plant injury, and soil microorganisms. Factors such as light, moisture, and temperature can all cause changes in plant metabolism and the rhizosphere effect. In summary, rhizosphere populations are dependent on many diverse interacting factors. In heavy textured soils oxygen become limiting, resulting in reduced rhizosphere populations compared to the coarse-textured soils.

## 5. Microorganisms and the outstanding problems

Modern and innovative approaches to solve the problems facing mankind of health, economy and food security rely mainly on living organisms that can be grown and breed easily in the lab. Great achievements in the biotechnology have been accomplished using microorganisms like bacteria, fungi and viruses. For example, *Rhizobium* has been used as bio-fertilizer, mycorhizae were used to promote the plant growth by uptake minerals from soil, *Bacillus thuringiensis* is to control the pest, Ti plasmid of *Agrobacterium* in genetic engineering and the field is opined for more molecules and compounds (proteins, enzymes, UV-absorbents, etc) extracted from those microorganisms that are adapted to survive the hostile environmental in the Gulf region. Also, in Qatar there have been a large number of bacteria strains that have been isolated from soils and being tested against different pathogens affecting economic plants (data not published). Such approach might be a good alternative to the classic one of using chemicals. *B. thuringiensis* strains are aerobic, Gram-positive found broadly distributed in the Qatari soil; produce a protein that has been used to control insect population. Another challenge to make clean environment in Qatar, bioremediation is being tested in Qatar to degrade and remove contaminants from wastewater and soil. Bioremediation processes rely mainly on the activity of those microorganisms to get rid of organic pollutants including polycyclic aromatic hydrocarbons (PAHs). This topic will be discussed latter with phytoremediation of contaminated soils and waters. In Qatar, information about the organic degrading bacteria from soil are scarce.

However, some recent serious works have identified some of these bacteria from different Qatari polluted soils using modern techniques [90, 91]. Bioremediation has obvious advantages over physicochemical remediation methods due to several merits: cost-effective, convenient, complete degradation of organic pollutants and no collateral destruction of the site material or its indigenous flora and fauna. Nutrients such as nitrogen and phosphorus can be added to the soil to improve the effectiveness of land farming alone. These nutrients can boost the population of naturally occurring microbes. Numerous soil bacteria, including *Pseudomonas* sp. have the ability to degrade organic contaminants so some remediation will occur. Even with this augmentation, however, larger and more recalcitrant compounds generally are not remediating at satisfactory rates. The concept of using plants for remediation of organic pollutants emerged a few decades ago with the recognition that plants were capable of metabolizing many toxic compounds. There are many studies suggesting the usefulness of plant-microbe systems in the bioremediation of residual chemicals [92-95]. The presence of Plant growth-promoting rhizobacteria (PGPR) in rhizosphere can enable plants to achieve high levels of biomass in contaminated soils despite extreme conditions. Generally, PGPR function in three different ways: (i) by synthesizing particular compounds for the plants, (ii) facilitating the uptake of certain nutrients from the environment and (iii) preventing the plants from diseases [96]. The common traits of growth promotion includes production or changes in the concentration of plant hormones such as auxin, gibberellins, cytokinins and ethylene [95]. Bioaugmentation is a method to improve degradation and enhance the transformation rate of xenobiotics by the seeding of specific microbes, able to degrade the xenobiotics of interest. Extensive degradation of petroleum pollutants generally is accomplished by mixed microbial populations, rather than single microbial species. Many microbes are described to have the genetic tools to mineralize recalcitrant pollutants such as PAHs, chlorinated aliphatics and aromatics, nitroaromatics and long-chain alkanes. These microbes can be wild-type isolates and genetically modified strains equipped with catabolic plasmids, containing the relevant degradation genes [93, 94].

### 5.1. Restoration of habitats

Many habitats are disappearing from several locations in the Gulf region in general and in the State of Qatar in particular, with the completion of constructions and urban development due to the great expansion caused by the establishment of infrastructure of the extraction and industry of oil and gas [4, 97]. Also, cconsiderable achievements being accomplished in Qatar in the development of land and establishment of roads, high ways, new buildings and many facilities as preparations to host major political, sport and social activities in the next decade. In fact, during the last decade and with the beginning of these developments, many reports and studies have warned the decision makers and scientists that the natural wildlife in the Gulf area is at real risk and facing serious threat due to the human and industrial activities [41, 46, 98]. Such threat and damage to the environment could be deepened if these activities and industrial constructions are not coupled by any studies on habitat destruction, fragmentation or disturbances, and restoration and conservation measures to maintain the natural habitats. There is fear that before knowledge is obtained on the flora and fauna and their ecophysiological aspects, these habitats will be lost. Therefore, some serious measures should be taken urgently including vegetation maps and monitoring exercises to document the state of the vegetation [99]. Such efforts should include ecophysiological aspects of the wild life, as prerequisite for ecological restoration [5, 100].

Looking at the changes in Doha, many habitats in this city are being demolished and / or threatened with the completion of construction of new buildings and facilities to comply with the great expansions in the industrial and urban sectors and responsibilities of hosting a number of activities in the coming years. These changes that are taking place at the coastal areas and inland as well; are lacking appropriate impact assessment and restoration plans [98]. One example of these changes is well demonstrated in a location near Qatar University, the details of changes were discussed in some reports and papers [41]. Restoration of endangered habitats is urgently required to maintain and sustain wild life since many scientists and decision makers in the State of Qatar are aware of the serious threats facing the natural habitats especially at the coastal line and other parts of inland areas [4, 97]. Such threats had been neglected for long period of times in the past, and there is growing worries about the diminishing natural sources of land and good quality water for agriculture and the scanty seasonal rains as well as all types of wild life. It would be very useful for the local authorities to take the initiatives to avoiding the consequences of putting the environment at real risk due to these changes in all sectors of life in the State of Qatar. Therefore, successful ecological restoration and maintaining healthy environment are based on the following main principles: (1) Information, (2) Problems, (3) Plans, (4) Solutions, (5) Monitoring.

## 5.2. Information

Successful restoration programs need information about the environment and wildlife which includes inventory of fauna, flora and microorganisms, their morphological characteristics and the mechanisms of adaptation. In Qatar, there has been cconsiderable information documented in a number of publications covering the above topics [2, 5, 6, 10, 101-106]. Recording and documenting the existing wild life and their ecophysiology in their natural habitats have been considered as a first important step toward conducting successful ecological restoration. For example, Ecology and Flora of Qatar, a monograph written by Batanouny [6], covered the ecological features of Qatar, the landforms, the prevailing climatic conditions, soils and resources and their effects on the plant life, description of the vegetation and the widespread plant communities, and the flora of Qatar which includes description of 301 species. Another monograph worth to be reported here prepared by Abulfatih and his colleagues [5]: Vegetation of Qatar, which recorded the state of the plant communities and ecosystems in the state of Qatar especially at the locations of oil and gas industry. Such efforts were considered essential for monitoring the future state of natural habitats and very valuable to biologists, environmentalists, naturalists, agriculturists as well as the decision makers and planners. Other published works, worth to be mentioned here, covered various topics of wild plants, horticultural plants, fungi and algae and their chemical constituents [2, 9, 98, 102, 104, 106-109].

During the last ten years some studies on the ecophysiology of wild plants in Qatar have shown that soil texture ranged between sandy to silty loam at the coastal line, while inland soils were sandy to sandy clay loam. The soils are dry, alkaline and highly saline, $EC_e$ ranged between 4 to 200 $dSm^{-1}$, and the most abundant ions in these soils and plants living in these soils are $Na^+$ and $Cl^-$ followed by $K^+$, $Ca^{2+}$, and $Mg^{2+}$. Most plant species are halophytes, undershrubs and succulents, with xerophytes due inland around Doha [3, 41, 46]. Halophytes did not accumulate much proline, soluble nitrogen and photosynthetic pigments, however, xerophytes do contain much of these organic components [4].

## 5.3. Problems

Recognition of the problems facing human life in all aspects, like economy, agriculture, health and wild life has been considered as the first step to successful scientific solution programs to tackle these problems. The problems facing the wild life in the State of Qatar can be summarized as follows: (1) harsh environment in terms of drought, salinity and high temperatures have great impact on the agricultural economy and the plant wild life. Changes in the climate, evaporation of water due to extreme high temperatures and scarcity of rainfall, and the intrusion of seawater into the underground water are the main reasons behind such problems. Irrigation of crops in the Gulf region faces serious challenges because of limited water supply of good quality suitable for normal plant growth. Moreover, large areas of the Qatari land are suitable for agricultural purposes; however, these areas are not only suffering from water shortage but also facing a problem of continuous increasing salanization. Soil salinity reached levels that inhibit the growth and yield of most crop plants, (2) disappearance of many coastal and inland habitats due to enormous urban constructions in various areas especially in Doha city which are accompanied with industrial expansions is putting the environment at risk, threatening ecosystem services and biological diversity, (3) wastewater accumulation at the outskirts of Doha and other towns due to the industrial activities of gas and oil could have great impact on the human and wild life. Pollution caused by heavy metals and organic hydrocarbons of the wastewater may cause real threats to many sectors of economy, agriculture, health and wild life.

## 5.4. Plans

Scientists have paid great deal of attention to the problems facing the mankind after considerable technological progresses to provide effective solutions to those problems. In the State of Qatar, there is a widespread perception among scientists and officials about the hazardous effects of pollutants resulted from accumulation of wastewater produced during gas and oil industrialization. Many serious measures have been taken in eliminating or reducing the adverse impact of those activities through a number of regulations to support scientific researches about the consequences of the loss of many natural habitats as well as the accumulation of organic and inorganic pollutants. For example, gas companies have pledged to the Supreme Council for Environmental Nature Reserve (SCENR) to implement Environmental Impact Assessment (EIA) for any plan to change the natural habitats. EIA is defined as a process of evaluating the impact and consequences of a project on various aspects of human life in both beneficial and adverse. Two main types of research to restore and conserve endangered habitats in Qatar can be conducted: (1) restoration of habitats, (2) bioremediation of contaminated soils, water and air. Various types of organisms can be used like bacteria, fungi and plants. A process of using plants to remove contaminants from soils is called phytoremediation; which will be discussed below.

### 5.5. Solutions

Solutions of the environmental problems lie always in hands of scientists, the decision makers and the public. Using scientific approaches to solve the environmental problems to minimize the consequences of the new developments in the Gulf region have been strongly suggested, active and serious efforts should start with the recording and documenting the existing wildlife including plants and their ecophysiology under their natural habitats. Such efforts can be considered as a prerequisite for successful ecological restoration [4, 9, 41, 98]. On the other hand, it would be very useful for the local authorities and the decision makers to take the initiatives to maintain the environment and solve the problems and easing the difficulties facing such efforts by: (1) implementing plans for restoration, (2) creating nature reserves to preserve the flora and fauna, (3) imposing laws that enforce the Environmental Impact Assessment (EIA) on oil and gas companies for any plans to change the natural habitats, (4) consulting research centers prior to any future construction plans, and, (5) activating scientific research for gene manipulation to maintain native plants. Halophytes and xerophytes living in these areas can be considered as good sources of salt-resistant and/or drought resistant traits from which genes can be transferred to crop plants [3, 9, 41, 97, 110]. The public, on the other hand, is the beneficiary or the victim of any environmental changes, and all the developments in all life sectors are reflected on this party. Therefore, to improve the awareness of conservation of natural habitats, educational materials should be enriched by various means like conferences, meetings, workshops, visits to some sites, as well as providing magazines, posters, pamphlets, films and videos.

### 5.6. Monitoring

Monitoring has been defined as collecting and analyzing of data to make the required and appropriate assessments of the performance success at various scales. In this concept such assessment might include modifications in the plans to conduct active amendments to solve the emerged problems and to build long-term public support to protect and restore habitats. All these efforts should be forwarded to the decision makers to take the appropriate measures. Thus, after the completion of environmental restoration, the establishment of long-term monitoring and resolving the arisen problems during that period. Supplement plans should be developed and ready for implementation to solve those emerged problems.

### 5.7. Case study of habitat restoration

The above five principles can be implemented in every successful ecological restoration plan, and the consecutive steps can be summarized as follows:

(A) Pre-Restoration process: (1) describe the current condition of ecological resources, (2) use the possible methods to describe the history of the site, (3) review the literature and visit the site to hypothesize how the original system of the site worked, (4) determine the objectives of the restoration and specify the condition of the site in the future.

(B) The Restoration process: (1) develop and implement the plan to achieve the objectives, this includes identification of the schedule tasks, adopt the methods, estimate the cost and labour, and begin the restoration work, (2) monitor the steps of the implementation plan to assess the success of restoration.

(C) Post-Restoration process: (1) prepare reports and evaluate the success periodically, (2) revise the plan and make the possible modifications to achieve the restoration objectives, (3) make education efforts to exploit the success of the restoration process.

In Qatar, little works have been done concerning the restoration of endangered habitats. However, a pioneer project to restore part of the coastal vegetation habitat was carried out during pipeline installation at Ras Laffan area. The details of this project can be found in the report of Al-Ansi and his colleagues [98], and was considered as a successful attempt to restore part of the coastal habitat. There are many reasons behind such success: (1) this project was a simple and inexpensive method to restore the vegetation, (2) the phases were implemented according to the plan, (3) in the pre-construction phase, the site was surveyed and the existing topography, vegetation density, floristic composition, and wildlife were recorded and documented in detail, (4) mitigation procedures were developed to minimize the impact of the construction activities in the project site, (5) tracking was undertaken to follow the progress of construction in the site, (6) after completing the construction, the project site was re-visited to ensure that the top soil was returned to its original position, and (7) the project site was inspected to monitor the restoration and growth of coastal vegetation.

## 6. Phytoremediation of contaminants

Contaminants enter into the environment in various ways, including direct leak or through accidents during transport or during waste disposal from storage sites or industrial facilities [111, 112]. Bioremediation process has been adopted as effective technique to remove, transfer, degrade and immobilize various types of pollutants from soil and ground water using living organisms including bacteria, algae and fungi. The term phytoremediation then used to refer to such process when plants and their associated microbes are used for environmental cleanup to deal with the organic and inorganic contaminants without the need to excavate them and dispose them elsewhere. The problems of contamination as a consequence of human and industrial activities are increasing. High levels of contaminants might have accumulated in the soil and urban areas near oil and gas facilities which cause a lot of damage to the environment and ecosystem. Most of the environmental contaminants are chemical of both organic and inorganic origins. Different phytoremediation mechanisms and methods have been recognized. The details of these processes can be found in many published reports and papers [113-117]. Phytoremediation is not a new technique to remove heavy metals or organic compounds from wastewater or contaminated soil. About 300 years ago some plants like *Thlaspi caerulescens* and *Viola calaminaria* were reported to accumulate high concentrations of heavy metals [118-120]. In the twentieth century, these efforts continued to add more plants that have the potential in removing toxic metals and organic compounds from soil and water. Many benefits have been reported of phytoremediation as compared to the conventional methods as follows: (1) less invasive and destructive, (2) less costs, (3) promote the biodiversity and enhance the restoration of the damaged habitats, (4) improve the environment components of water, soil and air, (5) reduce erosion by micrometeorological factors, (6) improve the general social life such as providing shade to buildings, decreasing energy consumption and reducing the carbon emitted from many sources [121].

Concerning the plant species growing in the State of Qatar that have been used in many studies of phytoremediation of pollutants from soils and waters elsewhere of the world included: (1) *Typha domingensis* Pers.: this plant has been used in phytoremediation studies to remove heavy metals from industrial wastewater and solution cultures [122, 123], (2) *Phragmites australis*: this species has been used in the phytoremediation of petroleum-polluted soils in China [11, 124], (3) Brassicaceae: members of this family are very important in phytoremediation of heavy metals [125-127], (4) *Juncus rigidus* Desf. members of the family Juncaceae might be used in the phytoremediation of contaminated soils [128], (5) *Tamarix* spp.: species of this genus had been used to produce wood by growing them in arid lands and irrigated them with salty effluent from desalinization plants or with recycled sewage [129], (6) *Prosopis Juliflora*: this tree can be used in phytoremediation of heavy metals [130, 131], (7) *Medicago* spp.: species of this genus like *Medicago sativa* have been used in phytoremediation of soil polluted with petroleum compounds [113]. Both species *Medicago laciniata* and *medicago polymorpha* are found within the flora of Qatar, and can be tested as an option in the phytoremediation, and (8) *Glycine max* has been successfully used in removing toxic petroleum products from contaminated soil [132]. More investigations are needed to increase the list of native plants that are efficient in removing pollutants from soil and water as expansion in the oil and gas industry and other human activities increased in the coming years. In Qatar, some serious research projects are being conducted to test some native plants to clean up soils and wastewater from organic and inorganic contaminants. Bioremediation (phytoremediation) processes have been considered as necessary and first step in successful ecological restoration of polluted habitats [133, 134].

## 7. Genetic approach

Since 1980 there have been serious and strenuous efforts by many scientists and researchers to develop crops with high resistance to the environmental stresses especially salinity and drought [135]. Since then great deal of achievements have been accomplished to identify many plants having some morphological, physiological and biochemical characteristics associated with the resistance to these abiotic stresses. Among these plants are wild plants and local varieties of some crops such as wheat, barley, rice and tomato [60, 136-144]. In Qatar and other Gulf States, native plants like halophytes and xerophytes deserve special attention, in addition to their economic and medicinal importance [106, 145], they could address important issues in biology and offer unique genetic pools to be used for gene technology programs leading to yield improvement of various crops under severe environmental stresses. Recent studies have concentrated on the native plants and to identify their critical salt or drought tolerance traits that could potentially be used in improving agricultural crops. Some reports [146] have compared *Arabidopsis thaliana* with 11 wild relatives in response to salinity. Major differences in some physiological traits including growth, water transport and ion accumulation were found, these differences can be exploited in the genetics of salt stress studies. Modern techniques that have been adopted to increase the resistance of crop plants to drought and salinity, including technologies of molecular biology, genetic engineering and tissue culture, have fruited in establishing solid basics to face the future challenges in the question of global food security [55, 110, 147-150].

Moreover, hopes are still in the minds of the decision makers and scientists to exploit the genetic bank in wild plants and / or genetically engineered plants to clean contaminated environments, and to deal with the pollution problems that arose due to the human activities and after the expansion in the industrial and urban sectors [151]. Striking successes have been achieved using genetic manipulation to improve the phytoremediation methods to remove pollutants from the environment as a step leading to the restoration of habitats [127]. Some reports [152] have indicated that using genetic engineering techniques is possible to improve some physiological characteristics in plants like uptake, transport, accumulation and tolerance of metals; such efforts could lead to create and develop transgenic plants have the ability to remove heavy metals from the growth medium. Thus, efforts using modern techniques and the identification of potentially genes for transformation of target plants could be promising approaches in improving the efficiency of these plants in the phytoremediation of contaminated environments. Some novel works worth to be mentioned here, some researchers [153] used chloroplast transformation to enhance the capacity of tobacco (*Nicotiana tabacum*) plant for mercury (Hg) phytoremediation, such technique may also have application to other metals that affect chloroplast function. Also, [128] have reported that wetland grasses and grass-like monocots can be changed genetically to improve their remediation potential. Plant species involved in these efforts are among those monocots genera in various families such as Poaceae, Cyperaceae, Juncaceae, and Typhaceae. Fulekar and his collaborators [154] have reported that plants such as *Populus angustifolia, Nicotiana tabacum* and *Silene cucubalis* have been genetically engineered to provide enhanced heavy metal accumulation characteristic as compared to the corresponding wild type plants. Other efforts in breeding plants having high biomass production and superior phytoremediation potential were considered as an alternative approach to deal with contaminants. The general productivity of plants is controlled by many genes, and genetic engineering techniques to implant more efficient accumulator gene into other plants have been suggested by many authors [115, 155]. Recently, some authors [156] in their review have concluded that transgenic plants and associated bacteria bring hope for a broader and more efficient application of phytoremediation for the treatment of organic compounds like polychlorinated biophenyls (PCBs). Genetic modification of plants may improve some phytoremediation mechanisms like phytoextraction, phytotransformation, etc, and also improve the bacterial efficiency in biodegradation of those organic compounds (rhizoremediation). The application of gene manipulation and the use of native plants that are metal tolerant and /or efficient in absorption and degradation of organic compounds should be accelerated and transferred from the experimental level to the field [157].

## 8. Conclusions

The ecophysiological studies of native plants that are adapted to various extreme environmental conditions like drought, salinity, high temperatures and contaminated environments are prerequisite to tackle the current problems facing mankind like food security, pollution and the endangered habitats. The State of Qatar and other Gulf States might have the preference over many other countries in the world to be leaders in the technological research to address the problems of restoration, phytoremediation and modern biological issues to deal with harsh environments. Being an Arab Gulf State, there are three basic elements to deal with the above problems: (1) the superior financial status which comes from the revenues of oil and gas industry, (2) the strong-willed of the scientists and decision makers. The universities in Qatar and other Gulf States and many established research centers have engaged and started actively with modern research to address various environmental problems, (3) the current native plants living in the Gulf States could be a good choice in the phytoremediation methods, and also might be good resources of traits from which genes can be manipulated and transferred to crop plants, or to develop efficient transgenic native plants in the phytoremediation processes of contaminated soil and water.

## Author details

Bassam T. Yasseen* and Roda F. Al-Thani

*Department of Biological and Environmental Sciences, College of Arts & Sciences, Qatar University, Doha, The State of Qatar*

## Acknowledgement

Authors would like to thank Professor Ekhlas M. Abdel-Bari, Environmental Studies Centre, Qatar University for providing the nice photographs of some wild plants from the Qatari habitats.

## 9. References

[1] Persian Gulf: http://www.emecs.or.jp/guidebook/eng/pdf/07persian.pdf. (accessed 28th September 2012).

[2] Abulfatih HA, Al-Thani RF, Al-Naimi IS, Swelleh JA, Elhag EA, Kardousha M. M., editors. Ecology of wastewater ponds in Qatar. Scientific and Applied Research Centre (SARC), University of Qatar, Doha, Qatar; 2002.

* Corresponding Author

[3] Yasseen BT, Al-Thani RF. Halophytes and associated properties of natural soils in the Doha area, Qatar. AEHMS 2007; 10: 320 – 326, DOI: 0.1080/14634980701519462.

[4] Yasseen BT. Urban Development Threatening Wild Plants in Doha City-Qatar: Ecophysiology is a Prerequisite for Ecological Restoration. Journal of Plant Sciences 2011; 6 (3): 113 – 123.

[5] Abulfatih HA, Abdel-Bari EM, Alsubaey A, Ibrahim YM., editors. Vegetation of Qatar. Scientific and Applied Research Center (SARC), University of Qatar, Doha, Qatar; 2001.

[6] Batanouny KH., editor. Ecology and Flora of Qatar. Scientific of Applied and Research Centre (SARC), University of Qatar, Doha, Qatar; 1981.

[7] Wikipedia, the free encyclopedia. Doha: http://en.wikipedia.org/wiki/Doha. (Accessed 28th September 2012).

[8] Meigs P., editor. World distribution of arid and semi-arid homo climates. In: Reviews of research on arid zone hydrology. Paris, Arid Zone Program, UNESCO; 1953. 1: p203 – 209.

[9] Abdel-Bari, EM, Yasseen BT, Al-Thani, RF. Halophytes in the State of Qatar. Environmental Studies Center. University of Qatar, Doha, Qatar, ISBN.99921-52-98-2; 2007.

[10] Norton J, Abdul Majid S, Allan D, Al Safran M, Böer B, Richer R., editors. An Illustrated Checklist of the Flora of Qatar. United Nations Educational, Scientific and Cultural Organization, UNESCO Office in Doha. Qatar Foundation. MAERSK OIL QATAR AS. Doha, Qatar; 2009.

[11] Obeid M. Qatar – study of the natural vegetation. FAO AGO QAT/74/003; 1975.

[12] Boulos L, editor.. 1978. Materials for a flora of Qatar. Webbia 1975; 32(2) :369-396.

[13] Turner NC. Drought resistance and adaptation to water deficit in crop plants. In: Mussel H, Staples RC. (eds.) Stress physiology in crop plants, Wiley and Sons (Interscience), New York; 1979. p343-372.

[14] Ludlow MM. Strategies in response to water stress. In: Kreeb HK, Richter H, Hinkley, TM. (eds.) Structural and functional response to environmental stresses: Water shortage. SPB Academic Press, The Netherlands; 1989. p. 269–281.

[15] Levitt J., editor. Responses of plants to environmental stresses. Vol. II. Water, Radiation, Salt, and Other Stresses. Academic Press, New York, London; 1980.

[16] Omara MK. Selection of early maturing barley with improved response to drought stress. Aust. J. Agric. Res. 1987; 38: 835-845.

[17] Chaves MM, maroco JP, Pereira JS. Understanding plant response to drought – from genes to the whole plant. Functional Plant Biology 2003; 30: 239 – 264.

[18] Zaman S, Padmesh S. Leaf anatomical adaptations of selected Kuwait's native desert plants. European Journal of Scientific Research (2009); 37 (2): 261 – 268.

[19] Yasseen BT., editor. Physiology of water stress in plants. The University Press, University of Mosul, Mosul, Iraq; 1992.

[20] Nilsen ET, Orcutt DM., editors. Physiology of plants under stress. Abiotic Factors. John Wiley & Sons, Inc., New York; 1996.

[21] Taiz L, Zeiger E., editors. Plant Physiology, 5th Ed. Sinauer Associates, Inc., Publishers, Sunderland, Massachusetts USA. ISBN 978-0-87893-866-7; 2010.

[22] Yasseen BT. An analysis of the effects of salinity on leaf growth in Mexican wheats, PhD thesis. The University of Leeds, Leeds, UK; 1983.

[23] Gulzar S, Ajmal Khan M, Ungar IA, Liu X. Influence of salinity on growth and osmotic relations of *Sporobolus ioclados*. Pak. J. Bot. 2005; 37 (1): 119 – 129.

[24] Gulzar S, Ajmal Khan M. Comparative salt tolerance of perennial grasses. In: Ajmal Khan, M, Weber, DJ. (eds.) Ecophysiology of high salinity tolerant plants. Springer, The Netherlands; 2006. p. 239 - 253.

[25] Brown GM., editor. Vegetation ecology and biodiversity of degraded desert area in north eastern Arabia. Kuwait University, Final report SO 073, Department of Biological Scienecs, Kuwait. pp. 15 – 16; 2001.

[26] Larcher W., editopr. Physiological plant ecology. Eco-physiology and stress physiology of functional groups, 4$^{th}$ ed. Springer, Berlin; 2003.

[27] Hoffman GJ, Shalhevet J, Meiri A. Leaf age and salinity influence on water relations of pepper leaves. Physiol. Plant. 1980; 48 (3): 463 - 469.

[28] Yasseen BT, Abu-Al-Basal MA, Alhadi FA. An analysis of leaf growth under osmotic stress. J. Plant Sciences 2010; 5: 391- 401.

[29] Byrne, K. (1998). Plant adaptation to dry environment. Bio Factsbeet. No. 29. http://www.scribd.com/doc/8303899/29-xerophytes. (Accessed 28th September 2012)

[30] Price AH, Cairns JE, Horton P, Jones HG, Griffiths H. Linking drought-resistance mechanisms to drought avoidance in upland rice using a QTL approach: progress and new opportunities to integrate stomatal and mesophyll responses. J. Exp. Bot. 2002; 53 (371): 989-1004.

[31] Wright GC, Smith RCG, McWilliam JR. Differences between two grain sorghum genotypes in adaptation to drought stress. I. Crop growth and yield responses. Aust. J. Agr. Res. 1983; 34: 615 – 626.

[32] Lee JS. Stomatal opening mechanism of CAM plants. J. Plant Biol. 2010; 53: 19-23.

[33] Koch KE, Kennedy RA. Crassulacean Acid Metabolism in the Succulent C4 Dicot, Portulaca oleracea L Under Natural Environmental Conditions. Plant Physiology 1982; 69: 757-761.

[34] Masrahi YS, Al-Huqail AA, Al-Turki TA, Sayed OH. Differential altitudinal distribution and diversity of plants with different photosynthetic pathways in arid southern Saudi Arabia. Australian Journal of Basic and Applied Sciences 2011; 5(6): 36-43.

[35] Mazen AMA. Crassulacean Acid Metabolism criteria shown by three species from flora of Qatar. QRS Repository 2011; Volume 2011; 2427. http://www.qscience.com/doi/abs/10.5339/qnrs.2011.2427.

[36] Mazen AMA. Changes in levels of phosphoenolpyruvate carboxylase with induction of Crassulacean acid metabolism (CAM)-like behaviour in the C4 plant *Portulaca oleracea*. Physiologia Plantarum1996; 98(1): 111-116.

[37] Clayton HAC, Barrett-Lennard EG, Ludwing M. Induction of CAM in the slender ice plant, Mesembryanthemum nodiflorum, by salinity and low water treatments. American Sciety of Plant Biologists. Photosynthesis & Respiration 2012; http://abstracts.aspb.org/pb2009/public/P45/P45025.html.

[38] Khan MA, Qaiser M. Halophytes of Pakistan: Characteristics, distribution and potential economic usages. In: Khan MA, Kust, GS, Barth HJ, Böer B. (eds.) Sabkha Ecosystems. Vol. II. Springer, Netherlands; 2006. p 129-153.

[39] Turner NC, O' Toole JC, Cruz RT, Yambao EB, Ahmad S, Namuco OS, Dingkuhn M. Response of seven diverse rice cultivars to water deficit. II. Osmotic adjustment. Leaf elasticity, leaf extension, leaf death, stomatal conductance and photosynthesis. Field Crop Res.1986; 13: 273 – 286.

[40] Dingkuhn M, Cruz RT, O'Toole JC, Turner NC, Doerffling K. Responses of seven diverse rice cultivars to water deficits. III. Accumulation of abscisic acid and proline in relation to leaf water-potential and osmotic adjustment. Field Crops Research 1991; 27 (1-2): 103 – 117.

[41] Yasseen BT, Abu-Al-Basal MA. Ecophysiology of Chenopodiaceae at the Coastline of Arabian Gulf-Qatar: Possible Destruction and Prespective Conservation. European Journal of Scientific Research 2010; 39 (1): 90 – 104.

[42] Aziz I, Gul B, Gulzar S, Khan MA. Seasonal variations in plant water status of four desert halophytes from semi-arid region of Karachi. Pak. J. Bot. 2011; 43(1): 587 – 594.

[43] Orcutt DM, Nilsen ET., editors. Physiology of plants under stress. Soil and Biotic Factors. John Wiley & sons, Inc. N.Y.; 2000.

[44] Aljuburi HJ, Al-masry HH. Effects of salinity and indole acetic acid on growth and mineral content of date palm seedlings. Fruits 2000; 55: 315 – 323.

[45] Ungar IA., editor. Ecophysiology of vascular halophytes. CRC Press, Boca Raton; 1991.

[46] Yasseen BT, Abu-Al-Basal MA. Ecophysiology of *Limonium axillare* and *Avicennia marina* from the coastline of Arabian Gulf-Qatar. Journal of Coastal Conservation: Planning and Management 2008; 12(1):35-42, DOI: 10:1007/s11852-008-0021-z.

[47] Glenn EP. Mechanisms of salt tolerance in higher plants. In: Basra AS, Nasra RK (eds) Mechanisms of environmental stress resistance in plants. Harwood Academic Publishers, Amsterdam; 1997. p 83 – 110.

[48] Joshi A J. Physiological studies on some halophytes. PhD thesis. Saurashtra University, India; 1979.

[49] Fahn A., editor. Plant anatomy, 3rd Ed. Pergamon Press, Oxford; 1990.

[50] Salama FM, El-Naggar SM, Ramadan T. Salt glands of some halophytes in Egypt. Phyton (Horn Austria) 1999; 39 (1): 91-105.

[51] Ahmed MZ, Gilani SA, Kikuchi A, Gulzar S, Ajmal Khan M, Watanabe KN. Population diversity of *Aeluropus lagopoides*: A potential cash crop for saline land. Pak. J. Bot. 2011; 43(1): 595-605.

[52] Longstreth DJ, Nobel, PS. Salinity effects on leaf anatomy. Consequences for photosynthesis. Plant Physiology 1979; 63 : 700 - 703.

[53] Albert R. Salt regulation in halophytes. Oecologia (Berl.) 1975; 21: 57-71.

[54] Flowers TJ, Yeo AR. Ion relations of plants under drought and salinity. Aust. J. Plant Physiol 1986; 13 : 75 – 91.

[55] Munns R, James RA, Läuchli A. Approaches to increasing the salt tolerance of wheat and other cereals. J. Exp. Bot. 2006; 57 (5): 1025-1043.

[56] Abu-Al-Basal MA, Yasseen BT. Changes in Growth Variables and Potassium Content in Leaves of Black Barley in Response to NaCl. Brazilian J. Plant Physiology 2009; 21 (4): 261 – 270.

[57] Wyn Jones RG, Storey R, Leigh RA, Ahmad N, Pollard A. A hypothesis on cytoplasmic osmoregulation. In: Marre E, O Cifferi O. (eds) Regulation of Cell Membrane Activities in Plants. Elsevier, Amsterdam; 1977. p 121-136.

[58] Greenway H, Munns R. Mechanisms of salt tolerance in nonhalophytes . Annu. Rev. Plant Physiol. 1980; 31 : 149 – 190.

[59] Flowers TJ. Physiology of halophytes. Plant & Soil 1985; 89: 41-56.

[60] Hasegawa PM, Bressan RA, Zhu J – K, Bohnert HJ. Plant cellular and molecular responses to high salinity. Annu. Rev. Plant Physiol. Plant Mol. Biol. 2000; 51: 463 – 499.

[61] Khan MA, Ungar IA, Showalter AM, Dewald HD. NaCl – induced accumulation of glycinebetaine in four subtropical halophytes from Pakistan. Physiol. Plant. 1998; 102 : 487 – 492.

[62] Sakamoto A, Murata N. Genetic engineering of glycinebetaine synthesis in plants: current status and implications for enhancement of stress tolerance. Journal of Experimental Botany 2000; 51: 81 – 88.

[63] Jolivet Y, Larher F, Hamelin J. Osmoregulation in halophytic higher plants: The protective effect of glycine betaine against the heat destabilization of membranes. Plant Sci. Letts 1982; 25: 193-201.

[64] Low PS. Molecular basis of the biological compatibility of nature's osmolytes. In: Eds. Gills R, Gilles-Baillien, M. (eds) Transport processes, iono-and osmoregulation. Springer-Verlag, Berlin; 1985. p 469-477.

[65] Xing W, Rajashekar CB. Glycine betaine involvement in freezing tolerance and water stress in *Arabidopsis thaliana*. Environmental & Experimental Botany 2001; 46 (1): 21 – 28.

[66] Chen TH-H, Murata N. Enhancement of tolerance to abiotic stress by metabolic engineering of betaines and other compatible solutes. Current Opinion in Plant Biology 2002; 5: 250–257.

[67] Park E-J, Jeknic Z, Pino, M-T, Murata N, Chen TH-H. Glycinebetaine accumulation is more effective in chloroplasts than in the cytosol for protecting transgenic tomato plants against abiotic stress. Plant, Cell & Environment 2007; 30: 994 – 1005.

[68] Chen TH-H, Murata N. Glycinebetaine protects plants against abiotic stress: mechanisms and biotechnological application. Plant, Cell & Environment 2011; 34 (1): 1-20.

[69] Montesinos E. Plant-associated microorganisms: a view from the scope of microbiology. Int. Microbiol. 2003; 6: 221–223. DOI 10.1007/s10123-003-0141-0.

[70] Barton LL, Northup DE., editor. Microbial Ecology. Wiley-Blackwell; 2011.

[71] Al-Whaibi MH. Desert Plants and Mycorrhizae (A mini-review). J. Pure Appli. Micro. 2009; 3(2): 457-466.

[72] Sigee DC., editor. Bacterial Plant Pathology; Cell and Molecular Aspects. Cambridge University Press; 1993.

[73] Mahasneh AM. Bacterial Decomposition of *Avicennia marina* L. Leaf Litter from Al-Khor (Qatar -Arabian Gulf). J. Biol. Sci. 2001; 2(11): 717-719.

[74] Mahasneh AM. Heterotrophic Marine Bacteria Attached to Leaves of Avicennia marina L. Along the Qatari Coast (Arabia Gulf) .J. Biol. Sci. 2002; 2(11)740-743.

[75] Holguin G, Vazquez P, Bashan Y. The role of sediment microorganisms in the productivity, conservation, and rehabilitation of the mangrove ecosystems: an overview. Biol. Fertil. Soils.2001; 33:265-278.

[76] Hoffmann L. Marine cyanobacteria in tropical regions: diversity and ecology. Eur. J. Phycol. 1999; 34(4): 371-379.

[77] Toledo G, Yoav Bashan Y, Al Soeldner. In vitro colonization and increase in nitrogen fixation of seedling roots of black mangrove inoculated by a filamentous cyanobacteria. *Cana. J. Micro.* 1995; 41(11): 1012-1020.

[78] Holguin G, Vazquez P., Bashan Y. The role of sediment microorganisms in the productivity, conservation, and rehabilitation of mangrove ecosystems: an overview.Biol. Fertil. Soils.2001; 33:265–278. DOI 0.1007/s003740000319.

[79] Rueda-Puente E, Castellanos T, Troyo-Diéguez E, Díaz de León-Álvarez J, Murillo-Amador B. Effects of a nitrogen-fixing indigenous bacterium *Klebsiell apneumoniae* on the growth and development of the halophyte *Salicornia bigelovii*as a new crop for saline environments. J. Agron. Crops Sci., 2003; 189: 323-332.

[80] Baskaran R, Vijayakumar R, Mohan P, M. Enrichment method for the isolation of bioactive actinomycetes from mangrove sediments of Andaman Islands, India. Malay. J. Micro. 2011; 7(1): 26-32.

[81] Gupta N, Das S, Basak, UC. Useful extracellular activity of bacteria isolated from Bhitarkanika mangrove ecosystem of Orissa coast. Malay J. of Micro. 2007; 3(2): 15-18.

[82] Ramesh S, Manivasagan P, Ashokkumar S, Rajaram G, Mayavu P. Plasmid Profiling and Multiple Antibiotic Resistance of Heterotrophic Bacteria Isolated from Muthupettai Mangrove Environment, Southeast Coast of India. Curr. Res. Bact. 2010; 3 (4): 227-237.

[83] Jalal KCA, NurFatin UT, Mardiana MA, Akbar John B, Kamaruzzaman YB, Shahbudin S, Nor Omar M. Antibiotic resistance microbes in tropical mangrovesediments in east coast peninsular, Malaysia. Afri. J. Micro. Res. 2010; 4 (8): 640-645.

[84] ] Fahmy GM, Al-Thani RF., editors. Ecology of Halophytes and their Bacterial Inhabitants in the Coastal Salt Marsh of Al-Dhakhira, Qatar.Environmental Studies Centre (ESC), University of Qatar, Doha, Qatar; 2006.

[85] Ouf SA. Mycological studies on the angiosperm root parasite *Cynomoriumcoccineum* L. and two of its halophytic hosts. Biologia Plantarum 1993; 35: 591-692.

[86] Brown AD. Microbial water stress. Bacteriolo. Rev.1976; 40: 803–846.

[87] Al-Thani RF. Microbiological Analysis of *Prosopis cineraria* (L.) Druce in the State of Qatar. In: The Ghaf Tree. *Prosopis cineraria* in Qatar. Abdel Bari EM, Fahmy GM, Al-Thani NJ, Al-Thani RF and Abdel Dayem MS (eds). Environmental Studies Centre at Qatar university and National Council for Culture, Arts and Heritage; 2007.

[88] Whitford W., editor. Ecology of Desert Systems.Academic Press. Oxford; 2002.

[89] Compant S, Duffy B, Nowak J, Cle´ment C, Barka E. Use of Plant Growth-Promoting Bacteria for Biocontrol of Plant Diseases: Principles, Mechanisms of Action, and Future Prospects. Appli. Env. Microbiol. 2005; 71(9): 4951-4959.

[90] Al-Thani R, Abd-El-Haleem D, Al- Shammri M. Isolation, Biochemical and Molecular Characterization of 2-chlorophenol Degrading *Bacillus* Isolates. Afric. J. Biotec. 2007; 6(23): 2675-2681.

[91] Al-Thani RF, Abd-El-Haleem DA, Al-Shammri M. Isolation and characterization of polyaromatic hydrocarbons-degrading bacteria from different Qatar soils. African J.Micro. Res.2009; 3(11): 761-766.

[92] Pieper DH, Reineke W. Engineering bacteria for bioremediation. Curr.Opin Biotec. 2000; 11(3): 262-270.

[93] Cohen MF, Yamasaki H, Mazzola M. Bioremediation of Soils by Plant–Microbe Systems Inter. J. Gree. 2004; 1(3): 301–312.

[94] Khan AG. Role of soil microbes in the rhizospheres of plants growing on trace metal contaminated soils in phytoremediation. J. Trace Elements in Med. Biol. 2005; 18 (4): 355-364.

[95] Zhuang X, Chen J, Shim H, Bai Z. New advances in plant growth-promoting rhizobacteria for bioremediation. Envi. Inter. 2007; 33(3): 406-413.

[96] Divya B, Kumar MD. Plant–Microbe Interaction with Enhanced Bioremediation. Res. J. Biotec. 2011; 6(4)72-79.

[97] Richer R. Conservation in Qatar: Impacts of Increasing Industrialization. Center for International and Regional Studies (CIRS), Georgetown University, School of Foreign Service in Qatar; 2008.

[98] Al-Ansi MA, Abdel-Bari EM, Yasseen BT, Al-Khayat JA. Coastal Restoration: Restoration of a coastal vegetation habitat at Ras Raffan industrial city. SARC, University of Qatar; 2004.

[99] Riegl B, Korrubel JL, Martin C. Mapping and monitoring of coral communities and their spatial patterns using a surface – based video method from a vessel. Bulletin of Marine Science 2001; 69 (2): 869 – 880.

[100] Böer B, Al-Hajiri S. The coastal and sabkha flora of Qatar: An introduction. In: Barth HJ, Böer B (eds.) Sabkha ecosystems vol. 1: the Arabian peninsula and adjacent countries. Tasks for vegetation science 36. Kluwer Academic Publishers, The Netherlands; 2002. p 63–70.

[101] Muftah AMA. Diniflagellates of Qatar water. PhD thesis. University of Wales, UK; 1991.

[102] Moubasher AH., editor. Soil fungi in Qatar and other Arab countries. The Scientific and Applied Research Council, University of Qatar, Qatar; 1993.

[103] Al-Ansi MA. Fisherirs of the State of Qatar. PhD thesis. University of Aberdeen, UK; 1995.

[104] Rizk AM, El-Ghazaly GA., editors. Medicinal and poisonous plants of Qatar. The Scientific and Applied Research Centre, University of Qatar, Qatar; 1995.

[105] Jaman SK, Meakins R., editors. Biodiversity of Animals in Kuwait. Centre for Research and Studies on Kuwait, Kuwait; 1998.

[106] Al-Easa HS, Rizk AM, Abdel-Bari EM., editors. Chemical constituents and nutritive values of range plants in Qatar. The Scientific and Applied Research Centre, University of Qatar, Qatar; 2003.

[107] Rizk AM, Al-Nowaihi AS., editors. The phytochemistry of the horticultural plants of Qatar. The Scientific and Applied Research Centre, University of Qatar, Doha, Qatar; 1989.

[108] Rizk AM, Al-Easa HS, Kornprobst JM., editors. The phytochemistry of the macro and blue-green algae of the Arabian Gulf. Faculty of Science, University of Qatar, Doha, Qatar; 1999.

[109] Abdel-Bari EM, Fahmy GM, Al-Thani NJ, Al-Thani RF, Abdel-Dayem MS., editors. The Ghaf tree, *Prosopis cineraria* in Qatar. Environmental Studies Centre, Qatar University, Qatar, ISBN 99921-58-88-3; 2007.

[110] Flowers TJ. Improving crop salt tolerance. J. of Exp. Bot. 2004; 55(396): 307– 319.

[111] Riser-Roberts E., editor. Bioremediation of Petroleum Contaminated Sites. CRC Press, Boca Raton, FL; 1992.

[112] Khan FI, Husain T, Hejazi R. An overview and analysis of site remediation echnologies. Journal of Environmental management 2004; 71: 95-122.

[113] Frick CM, Farrell RE, Germida, JJ. Assessment of Phytoremediation as an *In-Situ*Technique for Cleaning Oil-Contaminated Sites. Department of Soil Science University of Saskatchewan Saskatoon, SK Canada S7N 5A8, Submitted to: Petroleum Technology Alliance of Canada (PTAC) Calgary, A. 1999; http://www.clu-n.org/download/remed/phyassess.pdf.

[114] Vidali M. Bioremediation: An overview. Pur. Appl. Chem. 2001; 73 (7): 1163–1172.

[115] Ghosh M, Singh SP. A review on phytoremediation of heavy metals and utilization of ts byproducts. Applied Ecology and Environmental Research 2005; 3 (1): 1-18.

[116] Erakhrumen AA. Phytoremediation: an environmentally sound technology for pollution prevention, control and remediation in developing countries. Educational Research and Review 2007; 2 (7): 151-156.

[117] Jadia CD, Fulekar MH. Phytoremediation of heavy metals: Recent techniques. African ournal of Biotechnology 2009; 8 (6): 921-928.

[118] Baumann A. Das Verhalten von ZinksatzengegenPflanzen und im Boden. Landwirtsch.Vers.-Statn 1885; 31: 1-53.

[119] Hartman WJJr., editor. An evaluation of land treatment of municipal wastewater and physical siting of facility installations.Washington, DC, US Department of Army; 1975.

[120] Lasat MM. The use of plants for the removal of toxic metals from contaminated soil. Plant Physiology 1996; 118, 875-883.

[121] Van Epps A., editor. Phytoremediation of petroleum hydrocarbons. Environmental Careers Organization, U. S. Environmental Protection Agency Office of Solid Waste and Emergency Response Office of Superfund Remediation and Technology Innovation Washington, DC.; 2006.

[122] Carvalho KM, Martin DF. Removal of aqueous selenium by four aquatic plants. J. Aquat. Plant Manage. 2001, 39: 33 – 36.

[123] Hegazy AK, Abdel-Ghani NT, El-Chaghaby GA. Phytoremediation of industrial wastewater potentiality by *Typha domingensis*. Int. J. Environ. Sci. Tech. 2011; 8(3): 639 – 648.

[124] Nie M, Wang Y, Yu J, Xiao M, Jiang L, Yang J, Fang C, Chen J, Li B. Understanding Plant-Microbe Interactions for Phytoremediation of Petroleum-Polluted Soil. PLoS ONE 2011. 6(3): e17961. doi:10.1371/journal.pone.0017961.

[125] Baker AJM, Brooks RR. Terrestrial higher plants which hyperaccumulate metallic elements – A eview of their distribution, ecology and phytochemistry. Biorecovery 1989; 1 (2): 81–126.

[126] Ebbs SD, Lasat MM, Brady DJ, Cornish J, Gordon R, Kochian IV. Phytoextraction of cadmium and zinc from a contaminated soil. J Environ Qual. 1997; 26(5): 1424–1430.

[127] Lone MI, He Z-L, Stoffella P, Yang X-E. Phytoremediation of heavy metal polluted soils and water: Progresses and perspectives. J Zhejiang Univ Sci B. 2008; 9 (3): 210–220.

[128] Czakó M, Feng X, He Y, Liang D, Márton L. Genetic modification of wetland grasses or phytoremediation. Z Naturforsch C. 2005; 60(3-4): 285-91.

[129] Hegedus R, Koseros T, Gal D, Pekar F, Birone Oncsik M, Lakatos G. Potential phytoremediation function of energy plants (*Tamarix tetranda*pall. and *Salix viminalis*l.) n effluent treatment of an intensive fish farming system using geothermal water. ActaUniversitatisSapientiae, Agriculture and Environment. 2009; 1: 31 – 37.

[130] Prasad MNV. Phyto-products from *Prosopis juliflora* (Velvet mesquite) applied in Phytoremediation. Frontiers in Trace Elements Research and Education, 10th nternational Conference on the Biogeochemistry of Trace Elements, 13-18 July 2009, held in Chihuahua, Chih, Mexico, http://icobte2009.cimav.edu.mx. http://icobte2009.cimav.edu.mx

[131] Shukla OP, Juwarkar AA, Singh SK, Khan S, Rai UN. Growth responses and metal accumulation capabilities of woody plants during the phytoremediation of tannery sludge. Waste Manag. 2011; 31(1):115-23.

[132] Njoku KL, Akinola MO, Oboh BO. Phytoremediation of crude oil contaminated soil: The effect of growth of *Glycine max* on the physico-chemistry and crude oil contents of soil. Nature and Science 2009; 7 (10): 79 – 87.

[133] Mackova M, Dowling DN, Macek T. Phytoremediation and rhizoremediation, Theoretical Background (Focus on Biotechnology), Series Editors: Hofman M, Anné J. Springer, The Netherlands; 2006.

[134] Paz-Alberto AM, Sigua GC, Baui BG, Prudente JA. Phytoextraction of Lead-Contaminated Soil Using Vetivergrass (*Vetiveria zizanioides* L.), Cogongrass (*Imperata cylindrica* L.) and Carabaograss (*Paspalum conjugatum* L.). Env Sci Pollut Res. 2007; 14 7): 498–504.

[135] Epstein E, Norlyn JD, Rush DW, Kingsbury RW, Kelley DB, Cunningham GA, Wrona AF. Saline culture of crops : A genetic approach. Science 1980; 210: 399- 404.

[136] Norlyn JD. Breeding salt – tolerant crop plants. In: Rains DW, Valentine RC, Hollaender A. (eds.) Genetic engineering of osmoregulation, impact on plant productivity for food, chemicals and energy. Plenum pub. New York; 1980. p 293 – 309.

[137] Winter SR, Musick JT, Porter KB. Evaluation of screening techniques for breeding drought – resistant winter wheat. Crop Sci. 1988; 28: 512 – 516.

[138] Yasseen BT, Al-Omary SS. An analysis of the effects of water stress on leaf growth and yield of three barley cultivars. Irrig. Sci. 1994;14:157 – 162.

[139] Yasseen BT, Al-Maamari BKS. Further evaluation of the resistance of Black barley to water stress: preliminary assessment for selecting drought resistant barley. J. Agron. and Crop Sci.1995; 174: 9-19.

[140] Zhu J – K. Genetic analysis of plant salt tolerance using *Arabidopsis.* Plant Physiol. 2000; 124 : 941 – 948.
[141] Hawkesford MJ, Buchner P., editors. Molecular analysis of plant adaptation to the environment. Kluwer Academic Publishers, Dordrecht; 2001.
[142] Xiong L, Zhu J-K. Abiotic stress signal transduction in plants: Molecular and genetic perspectives. Physiol. Plant. 2001; 112: 152 – 166.
[143] Gopal J, Iwama K. *In vitro* screening of potato against water stress mediated through sorbitol and polyethylene glycol. Plant Cell Rep. 2007; 26: 693-700.
[144] Aazami MA, Torabi M, Jalili E. *In vitro* response of promising tomato genotypes for olerance to osmotic stress. African Journal of Biotechnology 2010; 9(26): 4014-4017.
[145] Saenger P., editor. Mangrove Ecology, Silverculture and Conservation. Kluwer Academic Publishers, Dordrecht; 2002.
[146] Orsini F, D'Urzo MP, Inan G, Serra S, Oh D-H, Mickelbart MV, Consiglio F, Li X, Jeong C, Yun D-J, Bohnert HJ, Bressan R A, Maggio A. A comparative study of salt tolerance parameters in 11 wild relatives of *Arabidopsis thaliana*. Journal of Experimental Botany 2010; 61(13): 3787–3798. doi:10.1093/jxb/erq188.
[147] Xiong L, Zhu J-K. Molecular and genetic aspects of plant responses to osmotic stress. Plant, Cell and Environment 2002; 25: 131–139.
[148] Yamaguchi T, Blumwald E. Developing salt-tolerant crop plants: challenges and opportunities. Trends in Plant Science 2005; 10 (12): 615-620.
[149] Arzani A. Improving salinity tolerance in crop plants: a biotechnological view. In Vitro Cellular and Developmental Biology-Plant 2008; 44 (5): 373-383.
[150] Ezawa S, Tada Y. Identification of *salt* tolerance *genes* from the mangrove plant Bruguiera gymnorhiza using Agrobacterium functional screening. Plant Science 2009; 176 2): 272-278.
[151] Flathman PE, Lanza GR. Phytoremediation: Current views on an emerging green echnology. J. Soil Contamination 1998; 7(4): 415-432.
[152] Kärenlampi S, Schat H, Vangronsveld J, Verkleij JAC, Lelie D. Genetic engineering in he improvement of plants for phytoremediation of metal polluted soils. Environ. Pollut. 2000; 107: 225-231.
[153] Ruiz ON, Hussein HS, Terry N, Daniell H. Phytoremediation of organomercurial compounds via chloroplast genetic engineering. Plant Physiol. 2003; 132: 1344–1352
[154] Fulekar MH, Singh A, Bhaduri AM. Genetic engineering strategies for enhancing phytoremediation of heavy metals. African Journal of Biotechnology 2009; 8 (4): 529-535
[155] Cunningham SD, Ow DW. Promises and prospects of phytoremediation. Plant Physiol. 1996; 110: 715-719.
[156] Van Aken B, Correa PA, Schnoor JL. Phytoremediation of Polychlorinated Biphenyls: New Trends and Promises. Environ Sci Technol. 2010; 44(8): 2767–2776. doi:10.1021/es902514d.
[157] Mudgal V, Madaan N, Mudgal A. Heavy metals in plants: phytoremediation: Plants used to remediate heavy metal pollution. Agriculture and Biology Journal of North America 2010; 1(1): 40-46.

# Chapter 4

# Scaling Up of Leaf Transpiration and Stomatal Conductance of *Eucalyptus grandis x Eucalyptus urophylla* in Response to Environmental Variables

Kelly Cristina Tonello and José Teixeira Filho

Additional information is available at the end of the chapter

http://dx.doi.org/10.5772/55288

## 1. Introduction

Estimates of water use by plants are becoming increasingly important to forest science. Researchers apply water use estimates to predict the control of canopy conductance and transpiration [14, 26, 46], where this information is useful to help troubleshooting the water resources management [37, 24, 32], the role of transpiration in native forests [3] and to quantify the demand for water in short rotation forests and in plantations of *Eucalyptus* sp [10, 40, 41].

The growth and development of plants is a consequence of several physiological processes controlled by environmental conditions and genetic characteristics of each plant species. Therefore, in order to better understand the growth, development and hydrological impact of a *Eucalyptus* plantation, it is necessary to know the factors that control water use. Great efforts come up in order to investigate the contribution of water balance components in the productivity of eucalyptus, with the need to integrate the effects of climate and management practices on the production of wood from planted forests of Eucalyptus.

Process-based models consist in evidence-based relationships, which necessarily contain a relation of cause and effect, whether physical or biological [42]. A fundamental aspect of ecological processes is that they are affected by spatial and temporal dimensions. In spatial terms, for example, measurements made on a leaf in terms of net primary productivity, can not be extrapolated directly to the tree's canopy , because for this extrapolation is necessary knowledge about the distribution of the canopy, the arrangement of leaves, availability of soil water and others. Likewise, the extrapolation to the forest and the ecosystem needs information previously dispensable in smaller scales. This notion of scales and their extrapolations are essential to avoid mistaken views and phenomena in a certain scale for larger or smaller scales. As the scale is broaden, more interactions occur between the growing number of compartments of the system, making it more difficult and laborious to study the cause-effect relationships from models based on processes.

Thus, this study aims to describe the ecophysiological behavior of *Eucalyptus grandis x Eucalyptus urophylla* at the leaf level in association with environmental variables in three stages of development, in order to provide subsidies for the development of models that can predict the ecophysiological responses of a lower scale and its extrapolation to a larger scale.

## 2. Determining the scales

For a description and quantification on the water flow of *Eucalyptus grandis x Eucalyptus urophylla,* in order to integrate information on the leaf scale from lower scales to a larger scale, the study was conducted in three different ages (scales) of different plant development, known as: pot, plot and watershed. For each scale, the seedlings of *Eucalyptus grandis x Eucalyptus urophylla* were produced by the method of mini-cuttings in plastic tubes in the clonal nursery. Within approximately 120 days, when they reached the conditions for dispatch to the field, were destined for planting in:

**Pot Scale:** The seedlings were transported to the experimental field of School Agricultural Engineering, University of Campinas - FEAGRI / UNICAMP and transplanted to pots of 100 $dm^3$ (Figure 1a). These pots had circular holes in their sides and bottom, in order to allow better root aeration and drainage of excess water. The substrate was composed of vermiculite, coconut fiber and rice hulls. The ecophysiological study began 120 days after pot planting and measurements of transpiration, stomatal conductance, leaf water potential, photosynthetic active radiation and atmospheric vapor pressure deficit were carried out from February 2007 to June 2008.

**Plot Scale:** The seedlings were transferred to the experimental area of FEAGRI/UNICAMP and planted at 3 x 2 m spacing forming a clonal population. The predominant soil of the experimental area is classified as typical dystroferric Hapludox [9]. The ecophysiological study began at 240 days after planting (Figure 1b) and to assist data acquisition it was necessary to implant a measurement tower of 3 meters in height disposed between the crop rows to reach the treetops. The ecophysiological monitoring of transpiration, stomatal conductance, leaf water potential, photosynthetic active radiation and atmospheric vapor pressure deficit was conducted from January to July 2008.

**Watershed Scale:** The study was conducted at Santa Marta Farm, located in the Igaratá, São Paulo State. The geomorphological division of the State of Sao Paulo, according to the Institute for Technological Research [16], the study area is located in the Atlantic plateau that is characterized as a highland region, consisting predominantly of Precambrian crystalline rocks, cut by basic intrusive and alkaline Mesozoic-Tertiary rocks. The relief of the watershed is called a relief of hills, dominated local amplitude 100-300 m and slopes of medium to high - above 15%, with high drainage density, closed to open valleys and alluvial plains inland restricted [16]. The soil of the plot of interest is the type Tb dystrophic Cambisol, Oxisols with clay. For monitoring the ecophysiological behavior, measurements were carried out in a stand of *Eucalyptus grandis* x *Eucalyptus urophylla*, 60 months after planting in 3 x 2 m spacing with the aid of a platform lift with a range of 18 meters in height (Figure 1c). Monitoring ecophysiological transpiration, stomatal conductance, leaf water potential, photosynthetic active radiation and vapor pressure deficit atmospheric were held throughout the month of August 2008.

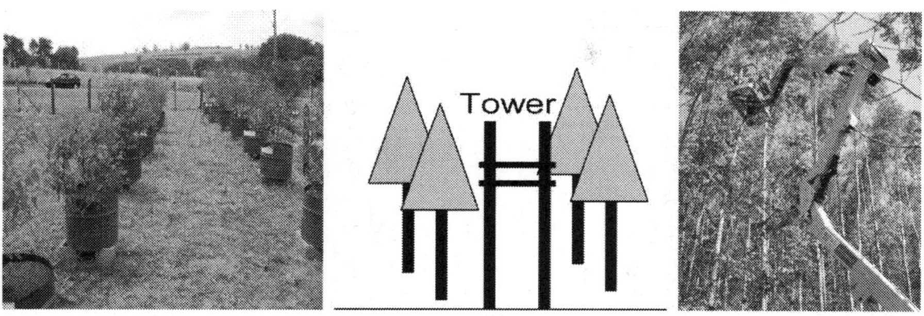


(a) (b) (c)

**Figure 1.** Scales: (a) Pot scale, (b) Plot scale and (c) Watershed scale: lift platform for observation of the variables used in ecophysiological clonal plantations of Santa Marta Farm, Igaratá-SP, Brazil.

## 2.1. Ecophysiological variables

The observations of water availability in the soil were performed by measuring predawn leaf water potential ($\Psi_{pd}$) using a Scholander pressure chamber [38], model 3035 (Soil Moisture Equipment Corp.., USA) before sunrise in healthy leaves fully expanded. According to [45], $\Psi_{pd}$ maintains a balanced relationship with the substrate's water potential, due to low rates of transpiration by plants presented overnight. To do so, four branches per seedling (on pot scale) or tree (on plot and watershed scales) were collected simultaneously. The measurements were carried out immediately after material collection.

Physiological measurements of transpiration (E) and stomatal conductance (Gs) were made by infrared gas analyzer (IRGA) LC-PRO + (ADC bioscientific Ltda., UK). For this end, four randomly and fully expanded healthy leaves for each individual seedlings/tree were chosen. The readings were held at hourly intervals throughout the day in the period from 8:00 am to 5:00 pm.

## 2.2. Environmental variables

Environmental variables such as photosynthetically active radiation and vapor pressure deficit of the atmosphere were chosen to correlate with the E and Gs. The PAR on the leaf surface (Qleaf) was determined simultaneously with measurements of ecophysiological variables, using the sensor coupled to the chamber of porometers, always disposed perpendicularly to incident sunlight on the leaf surface throughout each workday.

Additional data on air temperature and relative air humidity of the specific measurement days were obtained from an automatic weather station Campbell Scientific Inc. installed at the study site for each rating scale. This information was used to calculate the vapor pressure deficit of the atmosphere (VPD), as follows [30]:

$$VPD = es - ea, kPa \tag{1}$$

The saturation of vapor pressure (es) was calculated using the following equation:

$$es = 0.6108 * 10^{7.5*Tar\ /\ 237.3 + Tar}, kPa \tag{2}$$

Tar = air temperature, ° C

The partial vapor pressure (ea) was obtained by the following equation:

$$ea = RH * es / 100, kPa \tag{3}$$

RH = relative humidity of the place, %.

## 2.3. Ecophysiological models and Scaling up

*Structuring the ecophysiological model in the pot scale*

The scaling up of information measured on the pot scale was performed by applying the ecophysiological model used by [45] in order to simulate the E and Gs according to Qleaf, VPD and $\Psi_{pd}$ considering the hourly time scale of the period of study. Thus, follows the equation:

$$E = f(\Psi_{pd}, Qleaf, VPD) \tag{4}$$

$$Gs = f(\Psi_{pd}, Qleaf, VPD) \tag{5}$$

The models that relate the E and Gs (dependent variables) and environmental variables Qleaf VPD (independent variables) will be:

$$E = \beta_1 * Qleaf^2 + \beta_1' * Qleaf \tag{6}$$

$$Gs = \beta_2 * Qleaf^2 + \beta_2' * Qleaf \tag{7}$$

$$E = \beta_3 * VPD^2 + \beta_3' * VPD \tag{8}$$

$$Gs = \beta_4 * lnVPD + \beta_4' \tag{9}$$

Where: E - leaf transpiration (mmol $m^{-2} s^{-1}$); Gs - leaf stomatal conductance (mol $m^{-2} s^{-1}$); Qleaf - photosynthetic active radiation (μmol $m^{-2} s^{-1}$), VPD - vapor pressure deficit of the atmosphere (kPa), $\beta_1$, $\beta_2$, $\beta_3$, $\beta_4$ e $\beta_1'$, $\beta_2'$, $\beta_3'$, $\beta_4'$ = coefficients to be explained for each scale and model.

*Ecophysiological model calibration from pot to plot*

The adjustment of the ecophysiological model developed at the pot scale was given by the angular coefficient k, obtained by the ratio between the equations generated for the pot and plot scale, thus being:

- E: Scaling up from pot to plot ($E_P$ ')

$$k = E_P / E_V \tag{10}$$

$$E_P' = E_P * k \tag{11}$$

Where: Ev –equation *E= f (Qleaf or VPD)* on the pot scale; Ep - equation *E = f (Qleaf or VPD)* to plot scale; Ep'-adjusted equation of scaling up of *E = f (or Qleaf VPD)* from pot to plot by the angular coefficient of the model (k) to be specified for each scaling up.

- Gs: Scaling up from pot to plot (Gsp')

$$k = Gs_P / Gs_V \tag{12}$$

$$Gs_P' = Gs_P * k \tag{13}$$

Where: Gsv - equation *Gs = f (Qleaf or VPD)* on the pot scale, Gsp - equation *Gs = f (Qleaf or VPD)* to plot scale; $Gs_P$ '- adjusted equation of scaling up of *Gs = f (or Qleaf VPD)* from pot to plot by the angular coefficient of the model (k) to be specified for each scaling up.

The same methodology was adopted for the adjustment of the pot scale model to watershed scale ($E_B$' and $Gs_B$'') and plot to watershed ($E_B$'' and $Gs_B$''), being $E_B$ - equation *E = f (Qleaf* or *VPD)* and GSB - equation *Gs = f (Qleaf or VPD)* for watershed scale.

The ecophysiological model between the evaluation scales were subjected to analysis of variance and when significant, the means were compared by Tukey test using Minitab 14.0 software.

## 3. Case study: scaling up of the ecophysiological behavior of *Eucalyptus grandis x Eucalyptus urophylla* experience

Table 1 shows the daily mean of $\Psi_{pd}$ and environmental variables. The highest water comfort occurred on the plot scale, where we also observed a higher mean rate of transpiration. In assessing the environmental variables, note that the VPD situation between pot and plot, on average were similar, differing from the watershed scale that had mean lower than 33%, approximately. The energy available for physiological activity was higher in the pot scale, 100% higher than the watershed scale.

| Variable | Pot | Plot | Watershed |
|---|---|---|---|
| $\Psi_{pd}$ (MPa) | -0,30 | -0,15 | -0,21 |
| VPD (kPa) | 1,32 | 1,36 | 0,88 |
| Qleaf (μmol $m^{-2} s^{-1}$) | 1027 | 802 | 505 |

**Table 1.** Mean Ψpd and environmental variables for *Eucalyptus grandis x Eucalyptus urophylla* in the three scales.

### 3.1. Relation between E and Gs according to Qleaf and VPD

Figure 2 shows the interactions between gas exchange and environmental variables Qleaf and VPD. The values of E followed the evolution of Qleaf and VPD being almost imperceptible the difference between the curves of the plot and watershed scale. The observed values for the pot and watershed scales were between 0.9 and 11.4 mol $m^{-2} s^{-1}$ and for the plot scale, 1.3 to 13.3mol $m^{-2} s^{-1}$. The Gs also accompanied the increase in Qleaf, however, the greater tendency was found in the watershed scale. As for the VPD, the Gs showed smaller response amplitude in the plot and watershed scales, concentrated on the range of 0.1 to 0.5 mol $m^{-2} s^{-1}$. On the pot scale, it was observed the reduction of Gs with increasing VPD, with values close to 0.8mol $m^{-2} s^{-1}$ in situations of 1.0 kPa to 0.02mol $m^{-2} s^{-1}$ in extremes of VPD (3.0 kPa).

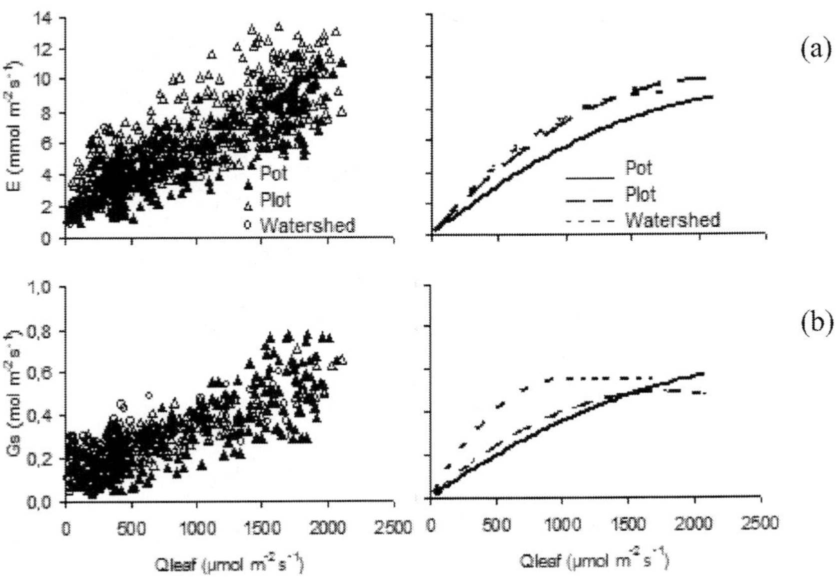

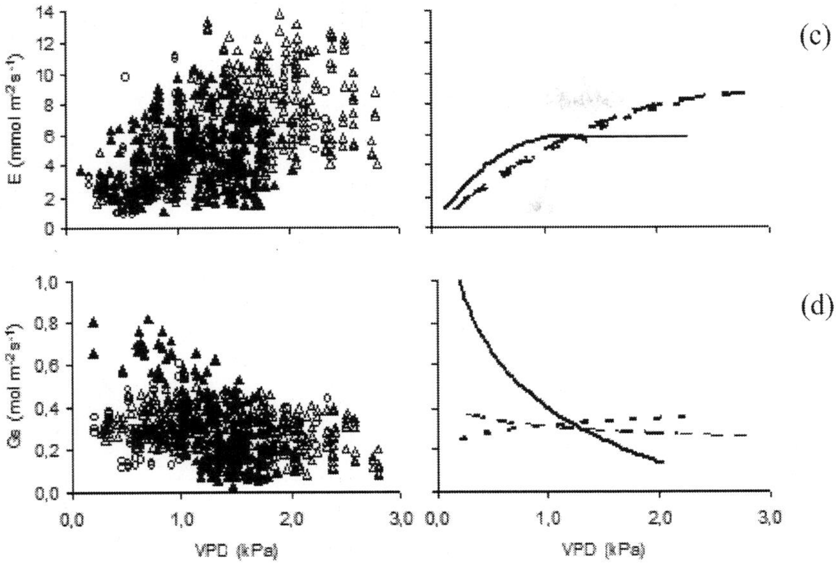


**Figure 2.** Behavior of E x Qleaf (a), Gs x Qleaf (b), E x VPD (c) and Gs x VPD (d) in the three scales of observation: pot, plot and watershed. Each point represents a mean of four measurements.

## 3.2. Interdependence of ecophysiological and environmental variables

In order to better understand the interdependence of ecophysiological variables (E and Gs) and environmental (Qleaf and VPD), it was established a ratio between the hourly mean of all values observed during the study period, E, Gs and their respective hourly mean of Qleaf and VPD (Table 2) With this mean ratio, we attempted to exclude the influence of variation of the concentration gradient of water and Qleaf or VPD in order to characterize the difference in diffusion behavior according only to the structure and physiology of the *Eucalyptus grandis x Eucalyptus urophylla*. The differences between the scales of study were significant in almost all majority. The exceptions were observed in the E / VPD between the plot and watershed scales that were similar (Figure 2b) and the ratio Gs / VPD between the pot and watershed scales, in contrast in this case, to what it was observed in Figure 2d.

| Between Scales | E/Qleaf | E/VPD | Gs/Qleaf | Gs/VPD |
|---|---|---|---|---|
| Pot x Plot | ** | ** | ** | * |
| Pot x Watershed | ** | ** | ** | ns |
| Plot x Watershed | ** | ns | ** | ** |

** e * = significant at 1% and 5% respectively, ns = non-significant

**Table 2.** Results of Tukey test comparing the means of the ratio E / Qleaf, E / VPD, Gs / Qleaf and Gs/ VPD for *Eucalyptus grandis x Eucalyptus urophylla* in the three scales of study.

Table 3 presents the correlation matrix between ecophysiological and environmental variables in the three scales of study. The best associations were present in the relations E and Gs with Qleaf, reinforcing the behavior displayed in Figure 2. On the plot scale, all correlations presented were significant and as in the pot scale, the relation between Gs and VPD was found to be negative, while at the watershed scale it was not consistent.

| Variables | Pot | Plot | Watershed |
|---|---|---|---|
| E x Qleaf | 0,86** | 0,83** | 0,87* |
| E x VPD | $0,30^{ns}$ | 0,63** | 0,56* |
| Gs x Qleaf | 0,80** | 0,85** | 0,66* |
| Gs x VPD | -0,76** | -0,33** | $0,07^{ns}$ |

** e * = significant correlation at 1% and 5% respectively, ns = non-significant

**Table 3.** Simple correlation matrix between ecophysiological and environmental variables on pot, plot and watershed scales.

### 3.3. Scaling up: pot, plot and watershed

Analyses of variance among the parameters were significant at 1% probability. The mathematical equations as well as the comparison between the hourly mean values observed and simulated by the models are in Tables 4, 5, 6 and 7. Among the relations, E was associated more evenly with Qleaf, with the highest coefficients of determination ($R^2$) when compared to VPD, regardless of the scale of observation. For this reason, Qleaf can be used more safely than other variables for being more consistent.

In model $E = f\ (Qleaf)$ the mean test was significant for $E_V$ and $E_B{'}$, indicating that the model of the pot scale, adjusted with k allowed extrapolation to the plot and watershed scales. Although the result of Table 2 has shown that the ratio E/Qleaf between plot and watershed scales are statistically different, it was not necessary to use k for the prediction of $E_B{''}$ (Table 2). The values observed and simulated by the models were compared and showed good correlation coefficients (Figures 3a, 3b and 3c), although the model underestimated the values in some situations (Figures 3b and 3c). As for the variable G, the model $Gs = f\ (Qleaf)$ could be applied in scaling up $Gs_P$ 'and $Gs_B{'}$ (Figure 4) not being significant for $Gs_B{''}$. The model $E = f\ (VPD)$ was adjusted for scaling up $E_{VP}$ and $E_{VB}$, but in the extrapolation of plot scale to watershed scale it was not necessary to adjust the constant k, reinforcing what was already observed in Figure 2 and Table 2. Although in this case the scaling up being possible, the correlation coefficients between the observed and simulated values by the model were between 0.66 and 0.62 and were highly significant ($P < 0.01$) (Figures 3d, 3e and 3f ). For $Gs = f\ (VPD)$, the proposed methodology can not be applied in any situation.

| Scale | n | $\beta_1$ | $\beta_1'$ | $R^2$ | k ± S.D. | $E_m$ ± S.D. | $Qleaf_m$ ± S.D. |
|---|---|---|---|---|---|---|---|
| Pot ($E_V$) | 199 | -0,000001 | 0,007 | 0,69 | - | 5,41 ± 2,07 a | 1027 ± 517 |
| Plot ($E_P$) | 516 | -0,000002 | 0,0094 | 0,64 | - | 5,60 ± 2,27 ab | 802 ± 465 |
| Watershed ($E_B$) | 78 | -0,000003 | 0,0106 | 0,83 | - | 3,96 ± 2,66 c | 505 ± 508 |
| Scaling up ($E_P'$) | 516 | -0,000001 | 0,007 | 1,00 | 1,25 ± 0,06 | 5,65 ± 2,62 b | 802 ± 465 |
| Scaling up ($E_B'$) | 78 | -0,000001 | 0,007 | 1,00 | 1,38 ± 0,14 | 3,56 ± 2,66 c | 505 ± 508 |
| Scaling up ($E_B''$) | 78 | -0,000002 | 0,0094 | 1,00 | - | 3,58 ± 3,12 c | 505 ± 508 |

Means followed by same small letter in columns do not differ by Tukey test at 5% probability. n = number of measurements (mean of 4 measures).

**Table 4.** Model coefficients of $E = f(Qleaf)$ with observed data from $E_v$, $E_P$, $E_B$, and adjustment to simulate the scaling up ($E_P'$,$E_B'$, $E_B''$), coefficient of determination ($R^2$), k, E and Qleaf mean ± mean standard deviation (k and $E_m$ ± S.D., mmol $m^{-2}$ $s^{-1}$; $Qleaf_m$ ± S.D., μmol $m^{-2}$ $s^{-1}$).

| Scale | n | $\beta_2$ | $\beta_2'$ | $R^2$ | k ± S.D. | $Gs_m$ ± S.D. | $Qleaf_m$ ± S.D. |
|---|---|---|---|---|---|---|---|
| Pot ($Gs_v$) | 199 | -0,00000009 | 0,004 | 0,64 | - | 0,34 ± 0,16 a | 1027 ± 517 |
| Plot ($Gs_P$) | 516 | -0,0000002 | 0,0006 | 0,59 | - | 0,28 ± 0,09 b | 802 ± 465 |
| Watershed ($Gs_B$) | 78 | -0,0000005 | 0,0011 | -0,17 | | 0,30 ± 0,09 bc | 505 ± 508 |
| Scaling up ($Gs_P'$) | 516 | -0,00000007 | 0,004 | 1,00 | 1,31 ± 0,12 | 0,26 ± 0,10 b | 802 ± 465 |
| Scaling up ($Gs_B'$) | 78 | -0,00000007 | 0,004 | 1,00 | 2,26 ± 0,53 | 0,26 ± 0,19 b | 505 ± 508 |
| Scaling up ($Gs_B''$) | 78 | -0,0000002 | 0,0006 | 1,00 | 1,43 ± 0,41 | 0,18 ± 0,11 d | 505 ± 508 |

Means followed by the same small letter in columns do not differ by Tukey test at 5% probability. n = number of measurements (mean of 4 measures).

**Table 5.** Model coefficients of $Gs = f(Qleaf)$ with observational data $Gs_V$, $Gs_P$, $Gs_B$ and adjustment to simulate scaling up ($Gs_P'$, $Gs_B'$, $Gs_B''$), coefficient of determination ($R^2$), k, Gs and Qleaf mean ± mean standard deviation(k and $Gs_m$ ± S.D., mol $m^{-2}$ $s^{-1}$; $Qleaf_m$ ± S.D., μmol $m^{-2}$ $s^{-1}$).

| Scale | n | $\beta_3$ | $\beta_3'$ | $R^2$ | k ± S.D. | $E_m$ ± S.D. | $VPD_m$ ± S.D. |
|---|---|---|---|---|---|---|---|
| Pot ($E_V$) | 225 | -0,4519 | 10,311 | 0,06 | - | 5,07 ± 1,87 a | 1,23 ± 0,35 |
| Plot ($E_P$) | 506 | -1,0627 | 6,0481 | 0,44 | - | 5,75 ± 2,32 b | 1,33 ± 0,49 |
| Watershed ($E_B$) | 78 | -0,927 | 5,6528 | 0,38 | - | 3,96 ± 2,62 c | 0,88 ± 0,41 |
| Scaling up ($E_P'$) | 506 | -0,4519 | 10,311 | 1,00 | 1,02 ± 1,24 | 5,79 ± 1,49 b | 1,33 ± 0,49 |
| Scaling up ($E_B'$) | 78 | -0,4519 | 10,311 | 1,00 | 0,57 ± 1,26 | 3,99 ± 1,47 c | 0,88 ± 0,41 |
| Scaling up ($E_B''$) | 78 | -1,0627 | 6,0481 | 1,00 | - | 4,19 ± 1,51 c | 0,88 ± 0,41 |

Means followed by the same small letter in columns do not differ by Tukey test at 5% probability. n = number of measurements (mean of 4 measures).

**Table 6.** Model coefficients of $E = f(VPD)$ with observed data from $E_V$, $E_P$, $E_B$ and adjustment to simulate scaling up ($E_P'$, $E_B'$, $E_B''$), coefficient of determination ($R^2$), k, E and VPD mean ± mean standard deviation (k and $E_m$ ± S.D., mmol $m^{-2}$ $s^{-1}$; $VPD_m$ ± S.D., kPa).

The scaling up with the involvement of the Qleaf in $E_B'$ and $E_B''$ generated values so similar that it is not possible to distinguish between these two models in Figures 3b and 3c. Similarly when using the variable VPD (Figure 3e, 3f).

| Scale | n | $\beta_4$ | $\beta_4'$ | $R^2$ | k ± S.D. | $Gs_m$ ± S.D. | $VPD_m$ ± S.D. |
|---|---|---|---|---|---|---|---|
| Pot ($Gs_V$) | 225 | -0,4015 | 0,4052 | 0,47 | - | 0,33 ± 0,19 a | 1,32 ± 030 |
| Plot ($Gs_P$) | 506 | -0,0505 | 0,3075 | 0,07 | - | 0,30 ± 0,08 b | 1,36 ± 0,51 |
| Watershed ($Gs_B$) | 78 | 0,046 | 0,3092 | 0,52 | - | 0,30 ± 0,10 b | 0,88 ± 0,41 |
| Scaling up ($Gs_{VP}$) | 506 | -0,4015 | 0,4052 | 1,00 | 3,85 ± 0,82 | 0,28 ± 0,01 c | 1,36 ± 0,51 |
| Scaling up ($Gs_{VB}$) | 78 | -0,4015 | 0,4052 | 1,00 | 0,94 ± 0,69 | 0,30 ± 0,02 d | 1,36 ± 0,51 |
| Scaling up ($Gs_{PB}$) | 78 | -0,0505 | 0,3075 | 1,00 | 0,93 ± 0,13 | 0,30 ± 0,02 e | 0,88 ± 0,41 |

Means followed by the same small letter in columns do not differ by Tukey test at 5% probability. n = number of measurements (mean of 4 measures).

**Table 7.** Model coefficients of *Gs = f (VPD)* with observational data $Gs_V$, $Gs_P$, $Gs_B$ and adjustment to simulate scaling up ($Gs_P'$, $Gs_B'$, $Gs_B''$), coefficient of determination ($R^2$), k, Gs and Qleaf mean ± mean standard deviation (k and $Gs_m$ ± S.D., mol $m^{-2}$ $s^{-1}$; $VPD_m$ ± S.D., kPa).

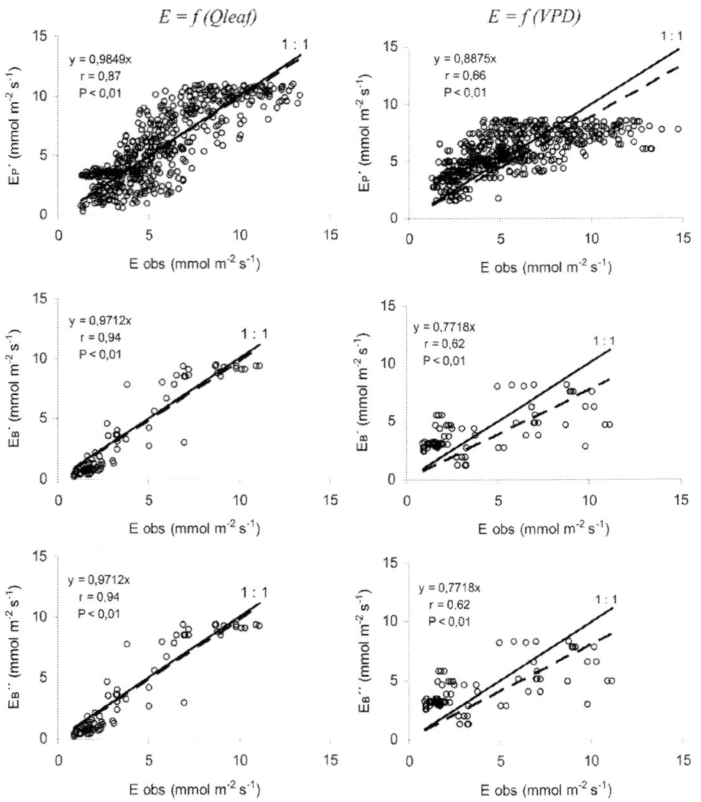


**Figure 3.** Linear regression (with intercept forced to zero) between the simulated values of E in the plot scale ($E_P'$) from the observed data in the pot scale ($E_{obs}$) according to Qleaf (a) and VPD (d), simulated values of E in the watershed scale ($E_B'$) from the observed data on the pot scale ($E_{obs}$) according to Qleaf (b) and VPD (e), simulated values of E in the watershed scale ($E_B''$) from the observed data in the plot scale ($E_{obs}$) according to Qleaf (c) and VPD (f).

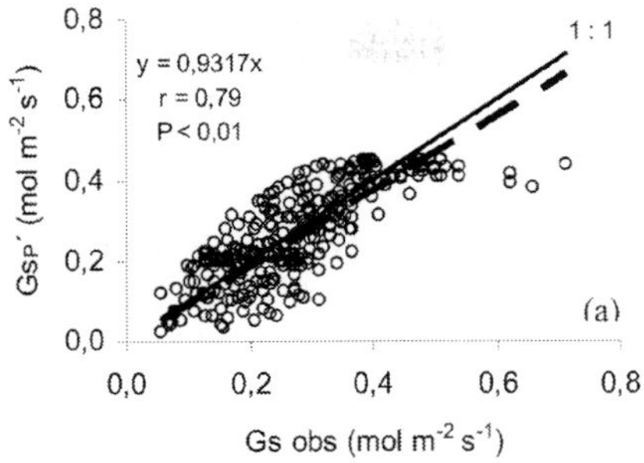


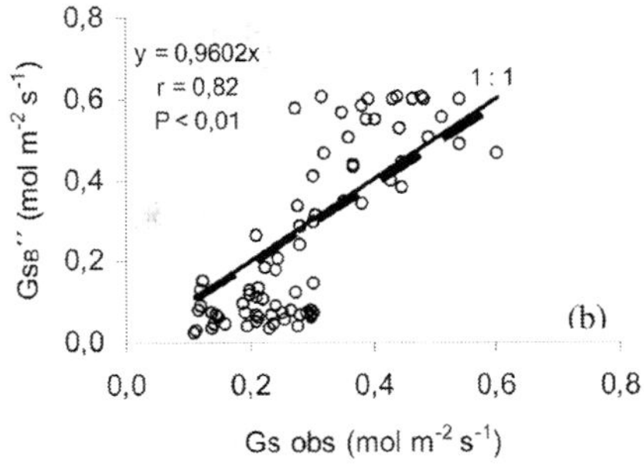


**Figure 4.** Linear regression (with intercept forced to zero) between the simulated values of Gs in the plot scale (GsP') from the observed data on the pot scale (Gs obs) (a), simulated values of Gs in watershed scale (GsB'') from observed data on the pot scale (Gs obs) (b) according to Qleaf.

## 4. Discussion

Many ecological studies are related to small spatial and temporal scales due to the easiness of operation and better understanding of the interaction of factors [18]. Considering this bias, the scaling up of information may constitute a useful tool for exploring upper scales from inferior ones and vice-versa [31]. This procedure involves a gradual process in which knowledge of how information is transferred from one scale to another is fundamental for understanding the mechanisms responsible for the natural generating of a standard phenomenon, which in turn are important for natural resource management. On the other hand, the scaling up is made from a reductionist perspective based on the detection mechanisms for determining the key processes operating at a certain level or scale, and its subsequent extrapolation to a higher or lower scale than the one studied [23].

From this principle, we could see the correlation between ecophysiological variables E and Gs with environmental variables Qleaf and VPD (Figure 2 and Table 3) in the three developmental stages of the *Eucalyptus grandis x Eucalyptus urophylla*. The characterization of the leaf behavior of E and Gs according to Qleaf and VPD from the pot scale to the watershed scale, we observed a similar tendency of response, which facilitated the extrapolation of the data from the pot scale to the plot/ or watershed scale in most relations, except for *Gs = f (VPD)*.

The behavior pattern of E and Gs according to Qleaf as observed in this study (Figure 2a, 2b, 2c), is well found in the literature [19, 22], as well as pattern of *Gs = f (VPD)* for the pot scale (Figure 2d). However, the almost linear response pattern of *Gs = f (VPD)* found in the plot and watershed scales showed lower Gs in these situations even with increasing VPD. This sharp difference in the tendency of response of *Gs = f (VPD)* between the scale pot with the plot and watershed scales led to difficulty in adjusting the equations $Gs_{VP}$, $Gs_{VB}$ and $Gs_{PB}$ (Table 7).

According to [39] and [48] there are numerous observations that Gs decreases in response to an increase in VPD between the leaf and air. However, the plot and watershed scales, where individuals were more mature than the pot scale, this behavior was somewhat modest, with little variation between the values of Gs. This fact can be justified by the findings obtained by [8] and [25], where they report that with the aging of individuals, the maximum levels of stomatal conductance decrease due to the greater sensitivity of stomata to the vapor pressure deficit of the atmosphere. Table 3 also shows the drop in correlation values between Gs and VPD with increasing age (scale) until appearing to be non-significant in the watershed scale.

In the other relations it was possible to predict the ecophysiological behavior with adjustment of the proposed model from young subjects, in this case, the pot scale with 120 days of age, for individuals with 240 days (plot scale) and for individuals with 60 months (watershed scale), approximately (Tables 4, 5 and 6). However, it also could be verified the accuracy of the model proposed for extrapolating the plot scale to the watershed scale in situations involving E, Qleaf and VPD without adjusting the model (Tables 4 and 6). There is a clear similarity between the ecophysiological responses of these two scales in Figures 2a, 2b and 2c.

The use of models seeks to simplify the complexity of real world privileging certain fundamental aspects of a system at the expense of detail. To provide an approximate view of reality, a model must be simple enough to be understood and used, and complex enough to represent the system under study [1]. The idea of proposing a model based on environmental variables (Qleaf and VPD) has been strengthened by [45], stating that the model reflects the conditions of the dynamics of the transport process in the soil-plant-atmosphere system, constituting the main component responsible for the flow of water in the plant.

Although the regression of E depending on Qleaf and VPD (Figure 2a and 2c) show the proximity of the tendency of the response of E between the plot and watershed scales with increasing Qleaf and / or VPD, the mean values obtained in field were lower for the watershed scale (Tables 4 and 6). This fact may be related to the reduced number of observations to the watershed scale compared to others, which may end up masking the results. Another important detail is related to environmental variables, both Qleaf and VPD that present, on average, lower than in the days of assessment of the pot and plot scales.

In [20] throughout the work "Physiology in forest models: history and the future" discusses the importance of understanding the operation and its ecophysiological approach in models of forest production. In literature, several papers are presented in order to relate the highest rates of gas exchange and growth of individuals or young forests with more mature ones. In [5] for example, discuss that the forest productivity increases during the rotation, reaches a peak near the period when the leaf area is maximum and then decreases substantially. But the reasons for this decline are not yet completely understood [33, 47]. The latest hypothesis about the decline in productivity with age was developed by [34], called the hydraulic limitation hypothesis.

As trees age, their hydraulic properties change, and at the same time, the amount of radiation intercepted by the canopy varies substantially [17]. With the increased size of the tree, water and nutrients must be transported over increasing distances between the root and the apex [7]. The water supply to the leaves of the apex becomes constrained by gravity and hydraulic conductance. These restrictions require a greater stomatal closure to maintain a minimum water potential to prevent xylem cavitation [15], resulting in decreases in gas exchange to a point where a positive carbon balance can not be achieved [4]. Some authors, by measuring transpiration by sap flow observed that the fall in productivity is accompanied by a decline in gas exchange rates [35, 2]. [28] studied the effects of age on the transpiration of a forest of *Pseudotsuga menziesii* of about 40 to 450 years of age in Oregon, USA, and by the sap flow methodology, also attributed to the hydraulic limitation hypothesis lower transpiration in individuals from the older forest, being that the highest transpiration of the 40 year-old forest provides further evidence of change in the local water balance because of its higher transpiration. In [2], on their turn, reported that the hydraulic limitation hypothesis proposes that the increased distance to be traveled by water inside the plant reduces the hydraulic conductance of the leaf. If the stoma closes to regulate the status of leaf water potential, taller trees will close their stomata at low vapor pressure deficits when compared to younger or shorter trees. Again, this report confirms the behavior observed in the plot and watershed scales to $Gs = f\,(VPD)$ (Fig. 2d, Table 3), however it was not the behavior observed for transpiration.

It is recognized that low pressure of water vapor between the leaf interior and the outside air (VPD) is an important environmental factor that affects the functioning of stomata. However, the causes for this event are still much discussed in the literature. The [44] examined the stomatal response to VPD in higher plants and the possible mechanisms proposed to explain this response. According to the author, the results are conflicting. When there is stomatal response to VPD, the mechanism that causes this response is also not well understood, being two hypotheses proposed for this mechanism. The hypothesis of "feedforward," which considers the decrease of Gs directly with increasing VPD, and abscisic acid (ABA), the signal for the response. In the event of feedback, Gs decreases with increasing VPD due to the increase in leaf transpiration, which lowers the water potential in the leaf. That is, the increase in E could be responsible for stomatal closure due to increased water potential gradient between guard cells and other epidermal cells or simply by reducing the leaf water potential [11, 27, 43]. These two mechanisms have been the subject of debate in the scientific community, for there are results published in the literature to support both hypotheses.

In any case, our results agree with the behavior explained by the hypothesis of feedback, even because we did not analyze the ABA during the study. [44] concludes his work as an unresolved issue, justifying the continuation of research in this area.

The hydraulic limitation hypothesis in some other studies failed to explain the reduced growth [2, 36] and the mechanism responsible for this fact was not identified. The [33] believes that there is no universal mechanism to explain the decline in productivity with increasing tree height, but that there are various components involved.

In searching for the characterization of the ecophysiological behavior of eucalyptus at different ages, [10] related to leaf area and rate of growth of *Eucalyptus globulus* Labill at the age of 2-8 years with stomatal conductance and transpiration by the method of sap flow in Australia. These authors observed an increase in transpiration of the stand from 2 to 5 years of age, where it reached to the peak in rates of exchange with subsequent decline thereafter. This decrease was related to the decline of leaf area index, with the result in annual growth rates and efficiency of water use. Although in our study, transpiration and stomatal conductance have been obtained at leaf scale by porometry, these variables had the same behavior found by them, i.e., the major tendencies of transpiration rates were observed in the plot and watershed scales, where individuals were more developed. We should also remember that the measurements, at whatever age (pot, plot or watershed scales), were always performed only in fully expanded leaves at the top of the canopy directly exposed to solar radiation. The difference between our study and [10] is that the evaluations performed by the method of sap flow are closely related to the total leaf area of the crown, without the need to quantify the leaves that consist it, nor the diversity in the degree of development of each one of them. The leaf area index is generally considered the most important determinant of differences in transpiration between different forest stands [13, 28]. Generally, young forests have a higher concentration of leaf area in a single layer of canopy, while as the tree grows, the leaves are more uniformly distributed in generating various vertical profiles of leaf area [29] and these changes in the distribution of stem and leaves can have pervasive effects on canopy transpiration.

The justification of this work for having higher gas exchange tendencies, at the leaf level , in the plot and watershed scales may be explained by the fact that individuals did not reach their peak of development, as justified by [10]. Thus, the physiological activities continue to "full steam" favoring the growth of biomass. So that, in terms of forest production, in the decision making about the best time for cutting the planted forest (*Eucalyptus* sp), it is studied the balance of production curves and mean and yearly increments, with the aim of identifying the maximum mean rate of increase in production. When this point is reached, it is said that this is the peak production of the forest, that is, when it reaches its greatest efficiency in production (technical age for cut-off). After this peak, there is a decline in the production curve, and economically speaking it is not feasible to keep the tree standing.

The scaling up of information held on a lower scale to a higher scale is more problematic for several reasons. The transpiration of most plant species, including eucalyptus, is determined by several factors that vary continuously. In addition to age, among them are climatic demand (solar radiation and vapor pressure deficit of the atmosphere, temperature and wind speed), the physiological mechanisms related to the stomatal response to environmental factors, water availability and soil nutrients [6, 20]. Another issue addressed by [12] which is normal to expect that the rate of perspiration varies from species to species, as well as vegetative growth.

Since transpiration is related to the development of leaf area in plantations of short duration such as eucalyptus, which have high rates of initial growth, can also happen fast maximization of water use by these crops, which ultimately generate implications for the prediction of its water needs and impacts on watershed hydrology.

A fundamental aspect of ecological processes is that they are affected by spatial and temporal dimensions. In spatial terms, for example, measurements made on a leaf in terms of net primary productivity, can not be extrapolated directly to the tree, because for this extrapolation it is necessary knowledge about the distribution of the canopy, the arrangement of leaves, availability of soil water etc. Likewise, the extrapolation to the forest and the ecosystem needs information previously dispensable in smaller scales. This study focused on all measurements, only fully expanded leaves and fully disposed to incident radiation. In fact, we know that there is no way to expand these results to an already formed canopy, since in this case, the leaves do not have a uniform development, as well as variation in the incidence of radiation and, thus, there is a need for additional data that were not addressed in this study. However, tendencies were observed and simulated by the models developed. This notion of scales and their extrapolations are essential to avoid mistaken views and phenomena in a certain scale to larger or smaller scales. As a scale is broadened, most interactions occur between the growing number of compartments of the system, making it more difficult and laborious studies of cause-effect relationships from models based on processes.

Adding to the complexity of understanding the interactions between the ecophysiological variables, there is also the difficulty of the experimental protocol. This fact reinforces the merits of the methodology presented here. As stated by [21] in order to generate practical tools, such calculations based on processes must be combined with empirical relations derived from experiments and measurements carried out over several periods.

## 5. Conclusions

The relations between E and Gs with Qleaf and VPD showed significant differences at 1 or 5% in all relations of scaling up, except for the relation E x VPD in scaling up of plot / watershed, and Gs x VPD pot / watershed. The relation that had the best response between E and Gs and environmental variables was E / Qleaf whose correlation was significant on all scales at 1 or 5%. The measured values of E and Gs were consistently above the plot and watershed scales compared to the pot scale. For each lower scale, a model was developed for scaling up into a higher scale. It was possible to perform the scaling up (pot, plot andwatershed scale) of E and Gs. The simulation models of E according to Qleaf / VPD and Gs and Qleaf / VPD proved robust in each of the scales. There was no need to adjust the models of scaling up between plot and watershed in relations involving E and Qleaf, and E and VPD. All results were obtained for $\Psi_{pd}$ between 0 to -0.5 MPa. It is suggested that measurements of E and Gs are carried out in three scales in others $\Psi_{pd}$ to confirm the findings of this study.

## Author details

Kelly Cristina Tonello
*Federal University of São Carlos, Departament of Environmental Science, Forest Engineering, Rod. João Leme dos Santos, km 110, Itinga, Sorocaba-SP, Brazil*

José Teixeira Filho
*Faculty of Agricultural Engineering, State University of Campinas (UNICAMP, Brazil), Cidade Universitária Zeferino Vaz, Campinas-SP, Brazil*

## 6. References

[1] Anderson M.G., Burt T.P. Hydrological Forecasting. John Wiley and Sons, 1985.

[2] Barnard H.R., Ryan M.G. A test of the hydraulic limitation hypothesis in fast-growing *Eucalyptus saligna*. Plant Cell Environ. 2003; v.26, 1235-1245.

[3] Barrett D.J., Hatton T.J., Ash J.E., Ball M.C. Transpiration by trees from contrasting forest types. Aust. J. Bot. 1996; v. 44, 249-263.

[4] Burgess S.S.O., Dawson T.E. Predicting the limits to tree height using statistical regressions of leaf traits. New Phytologist 2007; v.311, 1-11.

[5] Binkley D., Stape J.L., Ryan M.G., Barnard H.R., Fownes J. Age-related decline in forest ecosystem growth: an individual-tree, stand-structure hypothesis. Ecosystems 2002; v.5, 58-67.

[6] Calder I.R., Wright I.R., Murdiyarso D. A study of evapotranspiration from tropical rain forest – West Java. J. Hydrol. 1986; v.89, 13-31,

[7] Day M.E., Greenwood M.G., Diaz C. Age and size-relates trends in woody plant shoot developement: regulatory pathways and evidence for genetic control. Tree Physiol. 2002; v.22, 507-513,

[8] Domec J.C., Gartner B.L. Cavitation and water storage capacity in bole xylem segments of mature and young Douglas-fir trees. Trees 2011; v.15, 204-214.

[9] Embrapa. Sistema Brasileiro de Classificação de solos. Centro Nacional de Pesquisa de Solos; 1999.

[10] Forrester D.I., Collopy J.J., Morris J.D. Transpiration along an age series of *Eucalyptus globulus* plantations in southeastern Australia. Forest Ecol. Manag. 2009; doi:10.1016/j.foreco.2009.04.023.

[11] Friend, A.D. Use of a model of photosynthesis and leaf microenvironment to predict optimal stomatal conductance and leaf nitrogen partitioning. Plant Cell Environ.1991; v.14, n.6, 895-905.

[12] Gholz H.L., Lima W.P. The ecophysiological basis for productivity in the tropics. In.: Nambiar E.K.S., Brown A.G. (Eds.) Management of soil, nutrients and water in tropical plantation forests. ACIAR Monograph; 1997. 213-246.

[13] Hewlett J.D. Principles of forest hydrology. University of Georgia Press; 1982.

[14] Hinckley, T.M.; Richter, H.; Schulte, P.J. Water relations. In.: Raghavendra A.S. (ed) Physiology of trees. New York: John Wiley Sons; 1991. 137-162.

[15] Hubbard R.M., Ryan M.G., Stiller V., Sperry J.S. Stomatal conductance and photosynthesis vary linearly with plant hydraulic conductance in ponderosa pine. Plant Cell Environ. 2001.; v.24, 113-121.

[16] IPT - Instituto de Pesquisas Tecnológicas de São Paulo. Mapa geomorfológico do Estado de São Paulo. São Paulo, 1981.

[17] Irvine J., Law B.E., Kurpius M.R. Age-related changes in ecosystem struture and function and effects on water and crabon exchange in ponderosa pine. Tree Physiol. 2004; v.24, 753-763.

[18] Jarvis P.G. 1995. Scaling processes and problems. Plant Cell Environ., 18, 1079-1089.

[19] Jassal R.S., Black T.A., Spittlehouse D., Brümmer C., Nesic Z. Evapotranspiration and water use efficiency in different-aged Pacific Northwest Douglas-fir stands. Agric. Forest Meteorol. 2009; v.149, 1168-1178.

[20] Landsberg J. Physiology in forest models: history and the future. FBMIS 2003; v.1, 49-63.

[21] Landsberg J.J., Waring R.H. A generalised model of forest productivity using simplified concepts of radiotion-use-efficiengy, carbon balance and partioning. Forest Ecol. Manag. 1997; v.95, 209-228.

[22] Langensiepen M., Fuchs M., Bergamaschi H., Moreshet S., Cohen Y., Wolff P., Jutzi S.C., Cohen S., Rosa L.M.G., Li Y., Fricke T. Quantifying the uncertainties of transpiration calculations with Penman-Monteith equation under different climate and optimum water supply conditions. Agric. Forest Meteorol. 2009; v. 149, 1063-1072.

[23] Loehle C. Philosophical tools: potential contributions to ecology. **Oikos**1988; v.51, 97-104.

[24] Loustau D., Berbigier P., Roumagnac P., Pacheco-Arruda C., David J.S., Ferreira M.I., Pereira J.S., Tavares R. Transpiration of a 64-yearold maritime pine stand in Portugal. 1. Seasonal course of water flux through maritime pine. Oecologia 1996; v.107, 33-42.

[25] McDowell N.G., Phillips N., Lunch C., Bond B.J., Ryan M.G. An investigation of hydraulic limitation and compensation in large, old Douglas-fir trees. Tree Physiol. 2002; v.22, p.763-774.

[26] Meinzer F.C., Goldstein G., Jackson P.C., Holbrook N.M., Gutierrez M.V., Cavelier J. Environmental and physiological regulation of transpiration in tropical forest gap species : the influence of boundary layer and hydraulic properties. Oecologia 1995; v.101, 514-522.

[27] Monteith J.L. A reinterpretation of stomatal responses to humidity. Plant Cell Environ. 1995; v.18, n.2, 357-364.

[28] Moore G.W., Bond B.J., Jones J.A., Phillips N., Meinzer F. Structural and compositional controls on transpiration in 40- and 450-year-old riparian forests in western Oregon, USA. Tree Physiol. 2004; v.24, 481-491.

[29] Parker G.G., Davis M.M., Chapotin S.M. Canopy light transmittance in Douglas-fir–western hemlock stands. Tree Physiol. 2002; 22, 147–157.

[30] Pereira A., Angelocci L.R., Sentelhas P.C. Agrometeorologia fundamentos e aplicações. Guaíba: Agropecuária; 2002.

[31] Ramirez D.A., Bellot J., Domingo F., Blasco A. Stand transpiration os *Stipa tenacissima* grassland by sequential scaling and multi-source evapotranspiration modelling. J. Hydrol. 2007; v.342, 124-133.

[32] Roberts S., Vertessy R., Grayson R. Transpiration from *Eucalyptus sieberi* (L. Johnson) forests of different age. Forest Ecol. Manag. 2001; v.143, 153-161.

[33] Ryan M.G., Phillips N., Bond B.J. The hydraulic limitation hypothesis revisited. Plant Cell Environ. 2006; v.29, 367-381.

[34] Ryan M.G., Yoder B.J. Hydraulic limits to tree height and tree growth. Bioscience 1997; v.47, 235-242.

[35] Ryan M.G., Bond B.J., Law B.E., Hubbard R.M., Woodruff D., Cienciala E., Kucera J. Transpiration and whole-tree conductance in ponderosa pine trees of different heights. Oecologia 2000; v.124, 553-560.

[36] Ryan M.G., Binkley D., Fownes J.H., Giardina C.P., Senock R.S. An experimental test of the causes of forest growth decline with stand age. Ecological Monographs 2004; v.74, n.3, 393-414.

[37] Schiller G., Cohen Y. Water regime of a pine forest under a Mediterranean climate. Agric. Forest Meteorol. 1995; v.74, 181-193.

[38] Scholander P.F., Hammel H.T., Bradstreet E.D., Hemmingsen E.A. Sap pressure in vascular plants. Science 1965; v.148, 339-346.

[39] Schulze, E.D. Soil, water deficits and atmospheric humidity as environmental signals. In Water deficits: plant responses from cell to community. Smith J.A.C. and Griffiths H. Oxford: BIOS Scientific Publisher; 1993, 98-125.

[40] Soares J.V, Almeida A.C. Modeling de water balance and soil water fluxes in a fast growing *Eucalyptus* plantation in Brazil. J. Hydrol. 2001; v.253, 130-147.

[41] Stape J.L., Binkley D., Ryan M.G., Gomes A.N. Water use, water limitation, and water use efficiency in a *Eucalyptus* plantation. Bosque 2004a; v.25, n.2, 35-41.

[42] Stape J.L., Ryan M.G., Binkley D. Testing the utility of the 3-PG model growth of *Eucalyptus grandis* x *urophylla* with natural and manipulated supplies of water and nutrients. Forest Ecol. Manag. 2004b; 193, 219-234.

[43] Stewart D.W., Dwyer L.M. Stomatal response to plant water deficits. J. Theoret. Biol. 1983; London, v.104, n.3, 655-666.

[44] Streck N.A. Stomatal response to water vapor pressure deficit: an unsolved issue. R. Bras. Agroc. 2003; v.9, n. 4, 317-322.

[45] Teixeira Filho J., Damesin C., Rambal S., Joffre R. Retrieving leaf conductances from sap flows in a mixed Mediterranean woodland: a scaling exercise. Ann. Sci. Forestières 1998; v. 55, 173-190.

[46] Tognetti R., Giovannelli A., Lavini A., Morelli G., Fragnito F., D´Andria R. Assessing environmental controls over conductances trough the soil-plant-atmosphere continuum in an experimental olive tree plantation of southern Italy. Agric. Forest Meteorol. 2009; v.149, 1229-1243.

[47] Vanderklein D., Vilalta M., Lee S., Mencuccini M. Plant size, not age, regulates growth and gás exchange in grafted scots pine trees. Tree Physiol. 2007; v.27, 71-79.

[48] Yong J.W.H., Wong S.C., Farquhar G.D. Stomatal responses to changes in vapour pressure difference between the leaf and air. Plant, Cell and Environment 1997; v.20, 1213-1216.

Chapter 5

# Influence of the Root and Seed Traits on Tolerance to Abiotic Stress

Ladislav Bláha and Kateřina Pazderů

Additional information is available at the end of the chapter

http://dx.doi.org/10.5772/55656

## 1. Introduction

**Relationship between roots and seeds** is very important from the physiological point of view.

From the common view, the roots quality is modified by genotype and environmental conditions. But as follows especially from the wheat and other crops analysis the traits of roots are not only under genetic and environment control. Research results confirmed importance of individual seed traits for plant root growth and development, mainly seed vigor and subsequent seedling development at the poor conditions. These results show in all experiments statistically significant influence of abiotic stresses (drought, high temperature) on the seed traits and via the seeds influence on the traits of plant roots.

On the other hand influence of seed quality on the root growth and development of roots quality are basic assumption for the formation of high-quality seed of the most crops in next generation, especially at stress conditions. In the suboptimal conditions, the poor seed quality results in reduced root growth and also in low yield level. When the root growth from the beginning of vegetation is problematic, connected with the wrong roots quality, the negative consequences will continue during the whole vegetation period.

**Changes in seed germination during the year** exist in some species. Obtained results [20, 21] confirmed statistically significant relationship between speed of seed germination and intensity of geomagnetic activity during a year period.

**Long term seed storage conditions influence the following seedling growth** and the deterioration speed of the seed stored for a long time is affected by environmental conditions in which the seed was grown. From the practical point of view it is connected with question of the preservation of important genetic resources. Seeds from adverse environmental conditions rapidly lose germination energy and longevity. It can present a considerable economic costs connected with maintenance of genetic resources. This situation can lead even to loss of genetic resources.

Consequent changes during long term storage consist from: increasing concentration of free radicals, which are formed in time of long-term seed storage, damaging membrane lipids, inactivation of enzymes, damaging storage proteins and DNA. This process resulted to lower seed quality (low vigor) or even total loss of germination. Deterioration of the seeds during storage is irreversible phenomenon, natural for living organisms. Aged seeds influence optimal root growth (angle of root growth in the soil, depth penetration, tillering).

**Roots quality influences water utilization in plants** with different level by different cultivars in different environmental conditions and by this way drought tolerance during vegetation period and through the new filial seed generation germination and grow of young plants.

The water availability and efficiency of water utilization in time of germination is one of the basic factors influencing field emergence rate. Water uptake is the first step for enzymes activation, and shortly, for successful germination. The large variability in water use efficiency of seeds of different species and cultivars exist.

## 2. Seed and root phylogeny - general overview

The seeds, roots and their properties are the result of the phylogenic development under stress pressure, especially the influence of dry conditions in time of plant colonization of the Earth. **The seed phylogeny** reflects very interesting historical transition for photosynthetic organisms. The seed history consists of four main steps: the development of seed morphological structures, anatomy of seeds; the development of dormancy and the evolution of seed size (mass) [65, 66]

Seeds were developed over a period of approximately 300 million years of phylogeny for three main reasons [11, 14, 16, 63]

1. increase of species distribution area
2. preserve species for adverse conditions
3. enable efficient reproduction of species

The roots have from the paleontological view the first predecessors in –rhizoids- unicellular "fibres". [16]

So far, the oldest fossils of these plant organs - real roots came from the period 396 million years ago. As evidence for the findings of the first roots are two plants Rhynia and Sigillaria from the late Devonian period, where paleontological analysis revealed the depth of the roots with length no more than one meter. Plant roots are most sensitive part of plant body. The morphological and physiological root traits respond much more sensitive to the external environment than the aboveground parts of plants. Roots have a large share in the creation of soil, impact on the composition of the micro flora in the formation of ground, humus, the production of carbon dioxide, i.e. the composition of the atmosphere.

**Seeds and roots** have great importance for the abiotic stress tolerance during vegetation period (drought, high temperature). Seed traits determine plant growth on the beginning of vegetation period; especially by seed vigor, and by seed storage conditions. On the other hand the roots are influenced by seeds quality on the start of plant growth, especially root morphology, i.e. length, surface, deep of penetration of the roots and also root weight, number of root tips, number of root hairs, number of lateral roots and density of roots and modified by environmental conditions. Roots affect plants on the whole vegetation period and from this point of view influence growth and development of new seed generation. The level of theses relations depends on the environmental conditions; it means on the influence of seed provenance.

## 3. Short root and seed history step by step

When we skip the period 3.5 billion years ago, when the cyanobacteria was only on the Earth, the history of the plants can be divided into four periods:

1. Thalassiofyticum- (until 442 million years BC) period of *Algae*
2. Paleofyticum (442-248 million years BC)
3. Mesofyticum (248-97 million years BC)
4. Cenofyticum (97 million years BC until today)

In the first part of **Paleofyticum**, the ***"older" paleofyticum*** (442-354 million years BC) the plants spread from the water on the land. In this time formation of roots started, which enabled to live on the earth's surface. In the ***"younger" paleofyticum*** (354-298 million years BC) is characterized by seed development [16, 63].

The negative impact of environmental conditions -long period of the drought conditions- is reflected strongly in evolution. "Invention" of the seeds improved the chance of species survival thanks to the better dispersion of seeds and by this way expansion of the species. The evolution of plants is connected with ability to survive dry and cold period of plant growth. Some mechanisms were evolved, started with tissue specialization, through dormancy evolution to post harvest maturation.

In this period developed psilophyta, pteridophyta and plant ferns gymnosperms. Among the ferns (*Pteridophyta*) are classified plants lycopodium (*Lycopodiophyta*), horsetail (*Equisetophyta*), tree ferns (*Polypodiophyta*) and plants (Progymnospermophyta) as cycads and conifers.

**After transition of plants on the earth's surface** the influence of abiotic stressors prevail rather than biotic factors. Gradually, however, grew relationship among different species of the plants either negative or positive and between plants and other organisms

In **Mesofyticum** ancestors of the modern ferns appeared (fern seed plants, cycas, benetits, and conifers).

**Cenofyticum** is the last period of plant development, characterized by spreading of angiosperms. This period lasts until today. As examples there are given three examples of fossilized seeds (Figure 1, 2 and 3) History of seed development is from the physiological view very interesting [11, 14, and 33]

**Figure 1.** Fossilized seed, the recognizable is embryo, label (scutelllum), and endosperm. Czech Karst

**Figure 2.** Fragments of fosilised seeds, Czech Karst

**Figure 3.** Fossilized wallop, Czech Karst

## 4. Seed development and influence of the seed on the plant growth

The understanding of the relationship between seed development, environmental conditions and seed quality at the molecular, cellular, physiological and agronomical level are basic aim of seed science. On the beginning of seed formation is fusion of male and female gametes and double fertilization. After fertilization relatively short and quick seed development starts, i.e. seed become the primary recipient-sink for assimilates. This process means parallel growth and development of the seed, which includes initial cell formation, development of endoplasmic reticulum and growth of cell organelles – plastids, ribosome's, mitochondria and Golgi complex. After approximately three or four weeks of the seed development, starch and protein granules are main components of endosperm composition. Simultaneously with chemical changes, morphological, anatomical development and quick changes of seed weight occur. Every of these processes can be under influence of abiotic stresses and can be modified in a different way [85].

Development of the most seeds can be divided into three next phases [58]. The first phase is characterized by quick cells division and histodifferentiation to specific part of embryo (cotyledons, growing axis), together with storage tissues (usually endosperm). Next phase is growing phase (expansion) connected with reserves storage (carbohydrates, proteins, lipids) in storage tissues, mostly in cotyledons and endosperm. Whole process of seed development ends by decrease of water content (maturation phase), when seed metabolic activity is reduced and seed is passed to quiescence state (metabolic non-active).

All phases are different in water content at tissues of newly developed seed [59] (Water content increases in cell division phase together with total seed weight. In phase of cell elongation water content is stable, but dry matter of seed increase thanks to stored reserves. Quick decrease of water content characterizes maturation phase, till the level 10 – 15 % of water, specifically for each plant species.

Desiccation phase is typical for orthodox seeds (mostly plants from temperate climate) and necessary for overcoming adverse environmental conditions, evolved as adaptive mechanism in time of plant phylogeny. Thanks to lower water content orthodox seeds can be stored at normal conditions longer time without loss of theirs quality. This is very important exactly for seeds of plants used as agriculture crops.

Development of seeds and fruits is controlled by phytohormones [57]. After fertilization mainly cytokinins come to action to influence cell division in early stage of seed development. Gibberellins are connected with phase of reserves deposition, when keep endosperm in liquid state thanks to activation of $\alpha$-amylase [81, 75]. Auxins control cell elongation of newly formed seed. Influence of negative osmotic potential of surrounding tissues, which prevent the embryo to next development [89], is substituted with increase of absicic acid (ABA) level. ABA suppresses gibberellins' activity in the synthesis of □-amylase and by this mechanism prevents premature germination of new seed [13]. On the end phase ABA level decrease too, together with decrease of water, when the seeds are becoming quiescent. Regulation of ABA level in mature seed determines quiescent or dormant state of seed [79]. During the seed development on mother plant the environmental conditions can change and abiotic stress can appear. Seed development can be aborted in this case, when environmental conditions are very adverse. In early phase of development the embryo is very sensitive to lack of water [29]. Decrease of water potential more than above -1.6 MPa [73] can imply damage of embryo or other tissues. Less favorable environmental conditions can mean only formation of smaller seed, with smaller embryo or storage reserves, but smaller seed size is obviously associated with shorter seed survival [68].

Seed quality is affected by location of seed on mother plant and related with flux of assimilates to seeds. Each seed lot is heterogeneous group of seeds with similar characteristics thanks to the same origin. But if we evaluate each seed in detail, we can find differences between seeds.

Seed quality is influenced by sink: source ratio too. Buds, flowers and siliques of winter rape fall down more from lateral branches than from terminal, which is more preferred [62]. Similarly fruit size decline from margin to the centre of capitulum as the result of resource competition [86, 1]. Found that direct part of sunflower head is supplied from direct leaves. [5] confirm different content of oil, fatty acids and total tocopherol content in different seeds in sunflower head. We can suppose that similar differences exist between each seed in direct content of stored reserves and in synthesis and reduction of phytohormones and other metabolites. Stress conditions in time of seed development can these differences enhance.

All levels of adverse environmental conditions imply induction of stress response. Acclimation starts as the first step, when mechanism of protective compounds synthesis switches on, for example LEA proteins – dehydrins [32]. Generally scientists agree these proteins play important role in tolerance to water stress. Their protective function consists in stabilization of membrane structures [31, 84]. Angelovici [3] discussed gene expression and metabolic activation during desiccation of seed and their influence to the desiccation tolerance, dormancy competence and successful germination of the dry seeds. Together with genetically conditioned production of osmoprotective substances (as proline at rape tolerant to salinity [70] the maternal effect applies role in adaptation to environmental conditions too [46, 47, 88]. Dyer et al. [40] confirm the seed adaptation of some invasive species to adverse conditions in germination time on the mother plant in time of their maturation. They think that transgeneration plasticity (TGP) of seeds is the result only just seed adaptation on stressed plants. TGP can explain phenotypic move in adaptability of plants to worse environmental conditions and influence by this way more easy spread of species in environment.

The participation of environmental conditions on development of viable seed is generally taken to consideration; this influence can be even significant. When the stress conditions are continuous, developing seed can be damaged and on macroscopic level it means higher occurrence of less vigorous or even non-germinated seeds. In slightly adverse conditions protective compounds accumulate in seed and stay stored inside after desiccation. It has also been demonstrated that in seed stored rather than de-novo synthesized mRNAs play key roles during germination

The influence of seed on plant individual is formed in time of seed development on mother plant. Seed quality affects the germination process, which can be modified strongly by environmental conditions and next development of plant. The seeds germination is a complex physiological process comprising many metabolic pathways [66] with the goal to originate new plant as the next generation of plant species). Germination starts with uptake of water to the dry seed by imbibition, followed by metabolic changes in seed and ends with rupture of covering layers and emergence of radical protrusion. Figure 4 shows the example of the seed provenance on the root system of juvenile plants.

Ability of seeds to germinate in adverse environmental conditions is expressed as seed vigor can be explained as difference between germination percentages analyzed at optimal conditions by laboratory tests and between percentages of seedling emergence in field conditions (field emergence). There is lot of reasons for this difference (diseases, soil conditions, water content in soil, variability of temperature in the soil, etc.).

The influence of the seed provenance together with cultivar is very important. For example, the obtained results from the experiment with the organic and conventional seeds (four different provenances) of various spring cereal cultivars (bread and emmer wheat, *Triticum aestivum* L. and *T. dicoccum* Schrank, barley, *Hordeum vulgare* L. and oat, *Avena sativa* L.) confirmed importance of the cultivar and the seed provenance for the seed quality especially for the germination and efficiency of water utilization during this development phase. This is very important factor because of seed biological quality is one of basic factors, which has influence on the growth and development of the filial generation, especially in drought conditions [24] Good established crop stand is the basis for optimal development of agriculture plants to obtain next generation of seeds with high quality [76].On figure 5 and 6 are examples how the seed provenance influences the germination process. Comparison of wheat plants from different seed lots is on the figure **7.**

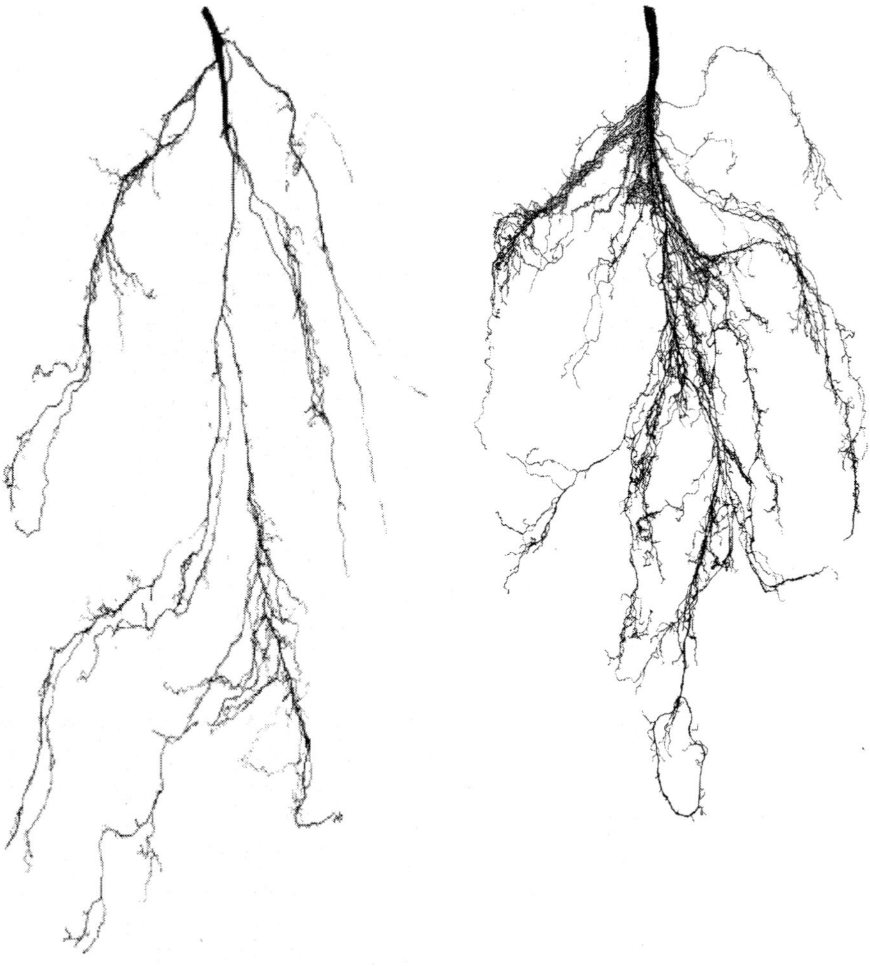

**Figure 4.** Example of seed provenance (variety Imari) effect on the root system at soya. On the left: from the seed from the dry conditions; on the right: seed from the standard environment. Influence of the seed provenance on the number of lateral root branches is evident

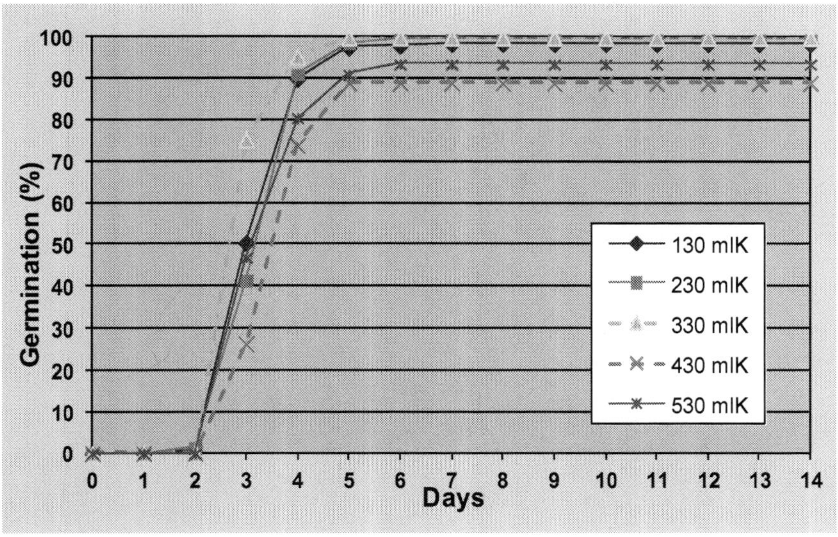


A

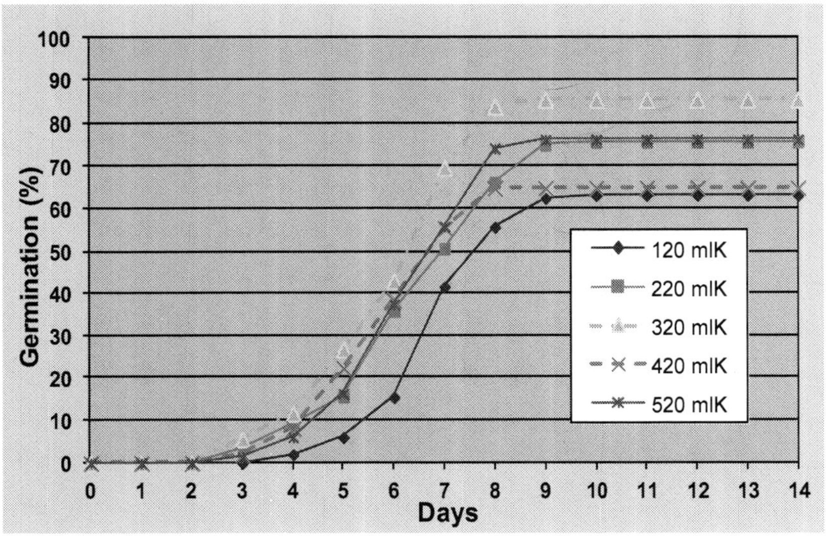


B

**Figure 5.** Difference between five spring barley seed lots germinated in standard (Fig A, filter paper, 30 ml of water, upper part) and in dry (Fig B, sand, 20 ml of water, lower part) conditions. Seed lots quality (expressed by germination curves and germination energy) looks very similar in optimal conditions, but can be very different in stress (drought) conditions. Difference of seed lot 3 (green) is evident.

**Figure 6.** Wheat plants grown in containers. On the left plants from the certified seeds obtained from the seed company; on the right two plants from different FSS (farmer save seed).

## 5. Importance of root traits for the seed growth and development

**Quality and physiological activity of the roots** is necessary condition for growth and optimal shoot development and subsequently development of the seeds with good quality at the majority of the field crops. This relation is also valid vice versa - the seed quality has positive influence on the root growth and development. Growing environmental conditions (=provenance), especially drought and high temperatures affect basic metabolic pathways, chemical composition of seeds, seed traits and traits of sprouting plants rootlets. These changes of the seeds are connected with efficiency of water utilization and especially with development of the root system. The research results show relatively large influence of seed traits on the root development. The genotypes with good germination under unfavorable conditions develop larger root system in field conditions, i.e. volume, length, deep of roots penetration after sowing and also during following vegetation period [17, 18, 19, 22, 28].

Coming climate change influencing mainly the variability of weather conditions during the vegetation period will increases the importance of the root system as the factor which plays more and more important role in the above-ground production and seed production too.

The importance of the seeds and roots are still neglected in the plant production. Misunderstanding of the roots importance can be found in the lot of scientific works, where detailed statistical analysis often evaluate only in above-ground parts of the plant physiological, anatomical, morphological and other properties, their relationships, but without knowledge about influence of the root system on the shoot analyzed traits and their relations.

Physiologically, the roots are the most sensitive part of the plant. As follows from the physiological literature, the basic changes of root system thanks to environmental conditions have following general way and influence: Drought changes deep of root penetration, low pH has influence on the length of roots, influence of salinity depends on the salinity type, high temperature influences number of root branches, low nutrients level is connected with the increase of the root system length, decrease of roots volume and number of branches, combination of abiotic stresses has large influence on the decrease of every trait, but not in every type of environmental conditions. These types of changes can have influence on transport of water and metabolites in plants and also on the shoot and seed growth and development. The developmental and growth stage, in which stress appear is very important. It is possible to show this on the rapeseed [17,18].

The speed of the root growth initialization after the winter period and the change of the ratio between dry weight of roots and above-ground parts in favor of the root mass in dry conditions during the vegetation period had the greatest impact on drought tolerance and yield of rapeseed. (The spring regeneration is expressed by the ratio of the dry weight of roots in spring and autumn. The variety with the largest growth rate of dry matter to roots has the seed production with best seed quality, i.e. weight, vigor, germination, optimal chemical composition).

**Darwin expressed that *"Roots are as brain of plants*"** [34], i.e. roots can be taken as a similar body like the brain. Currently, it is known that for the transmission signals (changes of potential) between root and above-ground plant parts plasmodesma are needed and there seems to be an important role for auxin molecule (IAA). For example, information about pathogen attack or strong physical stress can be quickly transmitted from the roots to the other plant parts in order to begin start as soon as possible organism defense thanks to plasmodesma. The genes for some plasmodesma proteins that form the connections are similar to the neuron proteins. The root system has the role as control centre with rapid transmission information to other plant parts. New situation creates a certain type of synapses, which are to some extent „the plants memory", i.e. certain type of reaction to already known situation. If the stress is repeated again, the reaction of the plant is more rapid on the basis of this memory [2, 4, 6, 7, 8, 9, 10, 11, 30, 56, 69, 80, 82, 83, 87, 91]. It is also known that root apices during growth can recognize in advance dangerous soil substrate and avoid them using similar active avoidance root tropism.

**Biochemical pathways - root influence of the shoot**

Last, but not least, there are significant advances in ecological studies, behavioral studies, on memory and learning phenomena in plants. **Baluška in his work gives a very detailed interpretation (verbatim quotation)** [6]. *"The plant neurobiological perspective reveals several surprises when the classical plant hormones like auxin, abscisic acid, ethylene, and salicylic acid are considered from this angle. Auxin and abscisic acid elicit immediate electric responses if applied to plant cells from outside, suggesting that their regulated release within plant tissues may be a part of neurotransmitter-like cell-to-cell communication. Abscisic acid signaling pathway is conserved between plants and animals and this signalling molecule both stimulates and is endogenously produced in human granulocytes in a way suggesting that it acts as endogenous proinflammatory cytokine. Biologically active abscisic acid was isolated also from brains of vertebrates indicating possible roles of abscisic acid in the central nervous system. Salicylic acid activates similar subset of MAPKs as voltage pulses. Ethylene, a classical plant hormonone, is an anaesthetic, a fact that plant physiologists have ignored until now. Interestingly, anaesthetics used on animals including man, induce anaesthetising effects on roots similar to those of ethylene. Ethylene is released in mechanically stressed plant tissues, and structurally diverse anaesthetics activate mechanosensitive channels. As ethylene is released after wounding, it might act to relieve 'pain' in plants. There are numerous other plant-derived substances, which manipulate the pain receptors in animals, such as capsaicin, menthol, camphor. Interestingly, the monoterpene volatiles, menthol and camphor induce oxidative stress and inhibit root growth in maize, indicating that they, too, act as plant signalling molecules. Finally, plants express inhibitors that are specific to the neuronal nitric oxide synthases. Another example of neuronal-like behaviour of plants is the report that prevention of nyctinastic movements of leguminous leaves causes their death while leaves allowed to 'sleep' stayed healthy. This resembles the situation in animals. Although melatonin was discovered in plants more than ten years ago, we know almost nothing about roles of melatonin in plants despite the fact that it is biochemically closely related to auxin. Interestingly in this respect, melatonin mimics auxin in the induction of lateral root primordia from pericycle cells."*

## 6. Influence of the seed traits on the root growth and development

### Introductory notes

Higher plants play the most important role in keeping a stable environment on the Earth to regulate global environment by many ways in terms of different levels molecular, individual, community, and so on [49,50], but the nature basis of the mechanism is gene expression and control temporally and spatially at the molecular level. There are many adverse stress conditions in the continuously changing environment, such as cold, drought, salinity and UV, which influence plant growth, development and crop and seed production.

Environmental conditions during the growing season affect biological quality of the evolving seeds, i.e. chemical composition of seeds, anatomical and morphological seed traits and traits of sprouting plants - basic metabolic seed pathways. Generally is known that the biological quality of seed is also one of basic factors, which influences growth and development of the roots at beginning of the filial generation and during vegetation period. It is possible to conclude that there is significant indirect influence of environmental conditions through the seed traits on the root development.

After harvest seed should be stored, i.e. there are storage conditions, which also affect the properties of seeds (enzyme activity, formation of free radicals).

**The seed traits and traits of sprouting plants** affect in filial generation especially root morphology at begin of vegetation period: length, surface, deep of roots penetration and also root weight, and later also number of root tips, number of root hairs, number of lateral roots and density of roots. Crop emergence, especially with good roots then influences the further course of growth. The start of period has significant influence on the following growth.

So far obtained results confirmed also in all experiments statistically significant influence of abiotic stresses (drought, high temperature) on the traits of seed and via the seeds influence on the root and shoot traits at begin of vegetation period. Very similar results among analyzed field crops in all types of experiments were obtained [17].

**This physiological phenomenon - germination (1) is influenced by the environment conditions during germination, during vegetation period and in time of new seed formation (2) and during new seed germination and growth (3). Resistance or tolerance to the environmental influences is (4) hereditary phenomenon.**

## 1. Germination

### Uptake of water

Uptake of water by mature dry seed is triphasic: I - rapid initial uptake (imbibition), II - plateau phase (metabolic processes) and III - this phase occurs when germination is completed. This phase is inhibited by the plant hormone ABA. The interactions between abscisic acid (ABA), gibberellins (GA) and brassinosteroids (BR) are regulating key processes that determine dormancy and germination. Abscisic acid inhibits germination and gibberellins and brassinosteroids promote germination.

**The first phase of germination** If the seed gets into the soil initially only water uptake exist, which start the first stage of germination, i.e. anaerobic processes, which is separated from embryo germination and for which there isn't need oxygen. This phase is mostly regulated by the amount of water in the seed, temperature and takes 24 -36 hours. Exactly imbibition phase is regulated only by differences in water potential between seed and soil. For example, in wet salted soil amount of water accessible for seed can be low, thanks to low water potential of soil than seed. Anaerobic respiration is predominant in this phase - the biological activity without oxygen (for example alcohol dehydrogenase activity).

The processes in the first phase can be affected by storage method, which affect the structural changes in the seed. Seeds of some varieties may then quickly absorb water, but in the case of drought loose of water is quicker too.

But transition from phase to phase is not sharp, for example embryo of maize seed can be highly hydrated and endosperm have low amount of water. On the other hand, activation of mRNAs in wheat seed was detected 2 hour after start of imbibition.

**The second phase of germination** is also under influence of the seed storage way, prolonged storage significantly affect mainly structural changes in the seed membranes and reduced enzyme activity. The manifestation of this changes may be, for example, decrease of germination energy (early germination) resulting in a field conditions in the low field emergence). In this phase begin - aerobic respiration system, citrate cycle, oxidative phosphorylation, and rapid activity of the mitochondria and lot of other biochemical processes. The visual germination process begins. What happens during germination? Metabolic activity in seed increases sharply, lipase, a-amylase, protease and peptidase - hence they are broken down starches, proteins, lipids, and their metabolic products are transported through the scutellum to the sprouting embryo – germination begins. Prerequisite for good germination is low ABA levels, which resulted in the loss of dormancy and increase the concentration of gibberellins thus promoting hormones, which creating a-amylase that breaks down starch.

Germination process according to the latest information also contributes brassinosteroids. Interactions between abscisic acid (ABA), brassinosteroids and gibberellins (GA) determine the level of dormancy and germination energy. These plant phytohormones (GA) with many functions in the plant during germination and seed sprouting are produced and throught label penetrates the endosperm to the aleurone layer, where it promotes the synthesis of $\alpha$-amylase, thus accelerating the starch breakdown. The ratio of starch, damaged starch, lipids and proteins like macro-elements can affect the germination rate.

Water uptake and efficiency of the water utilization are essential for enzyme activation, i.e. for use of reserve seed storage material [23, 24, 25, 26]. This trait has also influence on the root development at the begin of the vegetation period, but there is large variability in water use efficiency of seeds between different species, cultivars from different conditions (seed lots) and even between individual seeds from one seed lot [17, 19, 35, 36, 37, 38, 39, 40]. From the common view, the quality of root is modified by genotype, by environmental conditions and also by seed traits, especially thanks to seed vigor and young plants traits, especially at the poor conditions.

## 2. Stress during vegetation period

Water utilization by plants during vegetation period influences chemical composition of seed, but also traits of young plants in filial generation. The availability of water and efficiency of water utilization during germination is one of the basic factors that influence field emergence rate and following plantlet and plant root growth. The large variability of water use efficiency of seeds of different species and cultivars exists. This trait is under genetic and environmental control.

For each type of stress that acts on the developing seed the different changes in subsequent generation exist. So far obtained results confirmed that changes in the root system have resulted not only in changes in nutrient uptake, but may be reflected in the final stages of plant development and yield. In the case of nutrient uptake can be observed significant cultivars differences. The largest differences in nutrient uptake between the standard environment and stress environment (drought, high temperature, low pH) is previously in micronutrients (especially Zn, Mn, Fe), minor differences were then at macroelements (N, P, K, Ca, Mg). These changes may affect at more sensitive cultivars seeds properties.

In case of abiotic stress obtained results confirmed statistically significant influence of abiotic stresses at environmental conditions on the seed traits [27, 78]. These seed traits have a substantial effect on the tolerance to analyzed abiotic stresses in the filial generation and also on the root system and water utilization during germination. Similar results for spring and winter wheat were obtained [27, 78]. For example, severe stress during seed filling caused soybean plants to exceed their capacity to buffer seed number, shifting seed weight distributions towards a larger proportion of small seed, resulting in poor seed lot germination and vigor [39].

Stress during vegetation period for most crops changes anatomic structures of caryopsis. The most significant are changes in the layer of the pericarp, which creates a large number of cells which are different in size and shape. Seed anatomical changes are in accordance with the change of caryopsis morphology. Changing the aleurone layer structures can greatly influence the properties of all the economically important seed. This phenomenon highlights the importance of regularly exchange seed in crops production.

The influence of the environmental conditions changes energy content of the seeds. Grain influenced by abiotic stress is usually less vigorous compared with non-stressed plants. Seed vigor is also associated not only with weight, chemical composition, phytohormones activity, but also with change of the embryo properties. Negative influence of drought and high temperature conditions is reflected in the content of energy-rich substances accumulated in grain and straw, which are given not only the process of photosynthesis, by transport in plants, but also by the ratio and content of energy-rich substances, particularly sugars, proteins and fats.

High temperature stress of *Brassica napus* and other crops during flowering reduces micro and megagametophyte fertility, induces fruit abortion, and disrupts seed production [64].

## 3. Stress during new seed germination - seed stress tolerance

*It is known that it is possible to provide selection for cultivar resistance to stress already at the seed and at the seed germination stage and on the quality of the plant root system. Quality of the embryonic roots is important for the following growth and also roots development. In the juvenile phase and in later stages, these are the same genotype! This is a general biological regularity. At each experiment very good relationships between above named traits exist.*

For example it is possible to determine the genetic relationship between salt tolerance during seed germination and vegetative growth in tomato by comparing quantitative trait loci (QTLs) which confer salt tolerance at these two developmental stages. However, simultaneous improvement of tolerance at the two developmental stages should be possible through marker-assisted selection and breeding [43, 44].

Germination is also regulated by abscisic acid content. The content of abscisic acid in the seed is determined by genotype and conditions during the growth of seeds [78].

Signal transduction pathways, mediated by environmental and hormonal signals, regulate gene expression in seeds. Seed dormancy release and germination of species with coat dormancy is determined by the balance of forces between the growth potential of the embryo and the constraint exerted by the covering layers, e.g. testa and endosperm. GA releases dormancy, promotes germination and counteracts ABA effects. Ethylene and BR promote seed germination and also counteract ABA effects. We present an integrated view of the molecular genetics, physiology and biochemistry used to unravel how hormones control seed dormancy release and germination.

There are several ways to improve the adaptability of plants to the variable environmental stress conditions. Physiological studies of plant integrity have shown that the plant responds to stressors by modifying more than 100 physiological traits. The presented results [26] confirmed that seed vigor and plant vigor (quick escape from any stress) are in significant correlation with yield and root quality system.

Selected basic traits of seeds (vigor, germination percent, and emergence) and especially stress tolerance during germination of the seeds to the high and low temperature during day and night have significant influence on the quality of the root development. Plants with well-embryonic roots and high energy potential germination escape the stresses during begin the growing period, especially at drought conditions and are guarantee with high probability quality of the root system [64, 78,90].

## 4. Heredity, the availability of a selection at the level of seeds and seedlings

Contemporary knowledge confirm possibility of making selection for the root system and stress root tolerance on the basis of seedlings stress tolerance, i.e. at time of the sprouting. It is possible also to evaluate characteristics of seeds and seedlings, i.e. provide selection, after plant hybridization of the plants on the basis of the seed and seedlings traits for the seed quality an also for the classic selection in the plant breeding.

Seed of $BC_1$ progeny of an interspecific cross between a slow germinating *Lycopersicon esculentum* breeding line (NC84173; maternal and recurrent parent) and a fast germinating *L. pimpinellifolium* accession (LA722) were evaluated for germination under cold stress, salt stress and drought stress, and in each treatment the most rapidly germinating seeds (first 2%) were selected [43], .The results confirmed that rapid seed germination under a single stress environment may result in progeny with improved seed germination under a wide range of environmental conditions. Seeds of $F_2$ progeny of a cross between a slow-germinating (UCT5) and a fast-germinating tomato line (PI120256) were evaluated for germination under non-stress (control), cold-stress and salt-stress conditions, and in each treatment the most rapidly (first 5%) germinating seeds were selected, grown to maturity and self-pollinated to produce $F_3$ progeny.

In the case of phytohormones content that are genetically controlled by genes and it is not possible to draw any general conclusion about the correlation between the hormones content and the germination capacity during sprouting stress tolerance and water uptake. Why? Because all authors differ in their results and in conclusions; on the other hand it is possible to read that the fact that the sensitivity of the tissue towards hormones is also a very important factor in the development of regulation. High temperature stress during seed filling in controlled environments reduces soybean [*Glycine max* (L.) Merrill] seed germination and vigor, but the effect of high temperature in the field has not been determined. Contemporary findings support the results of experiments in controlled environments by demonstrating that high temperature during seed filling in the field, without seed infection with *P. longicolla* or physical injury, reduced soybean seed germination and vigor. Influence of the seed traits on the root system is known; especially at begin of vegetation period. Quality roots during the growing period are assumption for the creation of high-quality seed at most crops. *This relationship exists in reverse.* In the suboptimal conditions, the poor quality of seeds result in reduced growth and performance, quality and variety of the health of crops. When growth is at the beginning of vegetation period has the bad quality of embryonic roots according to bad seed quality in the suboptimal field conditions, the negative consequences are during all the vegetation period [27, 40, 48, 64, 71, 77, 78].

The root system can be affected by the quality of seeds especially at begin of vegetation period and change The worse seeds in stress environment can affect not only the quality of the root system with all the physiological consequences, *but also in some cases at some characteristics* subsequent generations, especially at the morphologic traits. If the combined effects of stressors during development and growth of seeds influence the subsequent generation through the seed traits it becomes especially at seedling traits and weight loss and mostly at the root system. This change is also at the chemical composition of the seeds. This phenomenon is still neglected in the plant breeding.

## 7. Conclusion

Development of the roots took place after the relocation of the plants to the surface of the Earth, i.e. long time before development of the seeds. The reason of the seeds development in the later time is to preserve the species, spread species and survive in unfavorable conditions (particularly by the development of dormancy). Importance of root traits for the seed growth and development is very significant and these relationships exist also in oposite direction - seed traits have influence on the root development. Seed quality is affected by location of seed on mother plant, by environmental conditions and by storage conditions.

The roots are, from the physiological view the most sensitive part of the plant. The root system has the role as control centre with rapid transmission information to other plant parts ("plant brain").

It is possible to provide selection for cultivar resistance to stress already at the seed germination stage and on the quality of the plant root system. Quality of the embryonic roots is important for the following growth and also roots development. In the juvenile phase and in later stage, there is the same genotype! This is a general biological regularity in nature.

It is also possible to evaluate characteristics of seeds and seedlings, i.e. make selection at this developmental phase, after plant hybridization on the basis of the seed and seedlings traits for the seed quality an also for the classic selection in the plant breeding.

## Author details

Ladislav Bláha
*Crop Research Institute, Division of Genetics and Plant Breeding, Prague, Czech Republic*

Kateřina Pazderů
*Czech University of Life Sciences, Department of Crop Production, Prague, Czech Republic*

## 8. References

[1] Alkio M; Diepenbrock W, Grimm E. Evidence for sectorial photoassimilate supply in the capitulum of sunflower (Helianthus annuus) DEC Springer New Phytologist 2002;156 (3) 445-456.

[2] Alpi A, et al. Plant neurobiology: no brain, no gain? Trends Plant Sci. 2007;(12):135–142

[3] Angelovici R, Galili G, Fernie A, R, Fait, A. Seed desiccation: a bridge between maturation and germination. Trends in plant science 2010 (4):211-218.

[4] Baluska F, Ninkovic V. (eds). Plant Communication from an Ecological Perspective. Series: Signaling and Communication in Plants, Vol. 6 Baluška, František; Ninkovic, Velemir (Eds.) ISBN 978-3-642-12161-62010

[5] Baluška F. Stress perception and adaptation in plant from neurobiology perspective in: Bláha L.edit Vliv abiotických a biotických stresorů na vlastnosti rostlin 2009;19-21.

[6] Baluška F, Mancuso S, Volkmann D. Communication in Plants: Neuronal Aspects of Plant Life. Springer-Verlag 2006.

[7] Baluška F, Mancuso S. Plant-Environment Interactions. Springer-Verlag; 2009.

[8] Baluška F. Volkmann D, Mancuso S. Communication in Plants: Neuronal Aspects of Plant Life. Springer Verlag: 2006.

[9] Baluška F. Plant Signaling. Springer-Verlag; 2009.

[10] Barlow PW. Biosystems. Reflections on 'plant neurobiology'. 2008;92(2)132-47.

[11] Baroux C, Spillane C, Grossniklaus U. Evolutionary origins of the endosperm in flowering plants. Genome Biology 2002;(3)1026.1–1026.5.

[12] Baydar H, Erbaş S. Influence of seed development and seed position on oil, fatty acids and total tocopherol contents in sunflower (Helianthus annuus L.) Turkish Journal of Agriculture and Forestry 2005; 29, (3) 2005,179-186.

[13] Bewley J. D, Black M. Seeds Physiology of Development and Germination. Plenum Press, New York. 1994

[14] Berger F. Endosperm, the crossroad of seed development. Current Opinion in Plant Biology 2003;(6)42–50.

[15] Black M, Corbineau F, Gee H, Come, D. Water content, rafinose, and dehydrins in the induction of dessication tolerance in immature wheat embryos. Plant Physiol, 1999,(120) 463-472.

[16] Bláha L.et al. Rostlina a stres. VÚRV, Praha, 2003, ISBN: 80-86555-32-1.

[17] Bláha L.Influece of seed quality on the root growth and development In :Proceeedings,7-th International symposium Structure and Functuion of Roots, Nový Smokovec, 5-9 september, High tatras, Slovakia, 2001

[18] Bláha L, Klíma M, Vyvadilová M.: The infleuence of the seed traits on the yield of selected genotypes of winter trape pieštany,In nové poznatky z genetiky a šlachťenija poľnohosspodárských rastlín,69-72Piešťany,November, 2011.

[19] Bláha L, Hnilicka F, Hořejší P, Novák V. Influence of abiotic stresses on the yield,seed and root traits at winter wheat (Tritium aestivum L.).Scientia Agriculturae Bohemica, 2003;34(1)1-7.

[20] Bláha, Gottwadová P. Klíčivost semen netradičních pícnin s rozdílnou proveniencí v různých stresových podmínkách Úroda 2008,(12) 36-42.

[21] Bláha L, Gottwaldová: Changes of seed germination during the year. Italian Journal of Agronomy: July September 2008; 3 (3) 387-388.

[22] Bláha L. Possibilities to Use Seed Traits of Grases for Drought Tolerance Prediction.In: Pazderu K. (ed.) Proceedings of 9th Scientific and Technical Seminar on Seed and Seedlings, CULS Prague, 2009;143-149.

[23] Bláha L, Hnilicka F, Kadlec P. Smrcková-Jankovká P, Macháčková I, Sychrová E, Kohout L. Influence of abiotic stresses on the winter wheat sprouting plants. Italian Journal of Agronomy, 2008, 3,(0), 389-390.

[24] Bláha L. Influence of the seed provenance on the germination and efficiency of water using In: Pazderu, K. (ed.) Proceedings of 10th Scientific and Technical Seminar on Seed and Seedlings, 2012;164-168,

[25] Bláha L, Hnilička F. Efficiency in water utilisation during seed germination. IN: Proceedings from Conference, Water productivity ina agriculture and horticulture: How can less water be used more efficiently 2-4 July, Viborg, Copenhagen 2007.

[26] Bláha L, Kadlec P, Kohout L, Gottwaldová P, Čepl J, Macháčkova I, Hnilička F. Vigour of seeds, quality of seed and influence of these traits on the selected crops, minor crops and potato for plant breeding, seed production ant plant production. In: Úroda12/2008; 53–60.

[27] Bouniols A, Texier V, Mondiès M, Piva G.: Soybean seed quality among genotypes and crop management : field experiment and model simulation. Eurosoya, 1998;(11) 87-99.

[28] Brenner E et al. Trends Plant Sci 2006;(11): 413-419.

[29] Berjak P, Vertucci C, W, Pammenter dessication-sensitive (recalcitrant) seeds: effects of developmental status and dehydration rate on charateristics of water and dessication-sensitivity in Camellia sinensis. Seed Sci Res, 1993;(3)155 -166.

[30] Brenner E, Stahlberg R, Mancuso S, Vivanco J, Baluška F, Van Volkenburgh E. Plant neurobiology: an integrated view of plant signaling. Trends Plant Sci. 2006;(11):413–419.

[31] Brini F, Hanin M, Lumbreras V, Amara I, Khoudi H, Hassairi A, Pages M, Masmoudi, K.: Overexpression of wheat dehydrin DHN-5 enhances tolerance to salt and osmotic stress in Arabidopsis thaliana. Plant Cell Rep,2007 (26)2017-2026.

[32] Close T,J. Dehydrins: a commonalty in the response of plants to dehydration and low temperature. Physiol Plantarum, 1997;(100) 291-296.

[33] Crane P,R, Herendeen P, Friis E,M. Fossils and plant phylogeny.American Journal of Botany 2004;(91) 1683–1699.

[34] Darwin C. The Power of Movements in Plants. London: 1880, John Murray;.

[35] D. B. Egli D.B, TeKrony M, Heitholt J.J, and Rupe J. Air Temperature During Seed Filling and Soybean Seed Germination and Vigor.Crop sci. 200545(4),1329-1335.

[36] Deno N,C. Seed germination theory and practice. Published by Norman C. Deno, 139 Lenor Drive, State College PA 16801, USA, second edition, 1993.

[37] Deno N,C. Seed germination theory and practice - first supplement. Published by Norman C. Deno, 139 Lenor Drive, State College PA 16801, USA, 1996.

[38] Deno N,C. Seed germination theory and practice - second supplement. Published by Norman C. Deno, 139 Lenor Drive, State College PA 16801, USA,1998.

[39] Dornbos Jr. L, Mullen R E: Influence of stress during soybean seed fill on seed weight, germination, and seedling growth rate Canadian Journal of Plant Sciences 1991, 71(2): 373-383

[40] Dyer A,R, Brown C,S, Espeland E,K, McKay J,K, Meimberg H, Rice K,J. The role of adaptive trans-generational plasticity in biological invasion of plants. Evol Appl, 2010(3)79-192.

[41] Egli D,B, *TeKrony D,M, Heitholt J, Rupe J. Air Temperature During Seed Filling and Soybean Seed Germination and Vigor.Crop.sci.2004° 45(4) 1329-1335.

[42] Ellis R. Rice seed quality development and temperature during late development and maturation. Seed Sci Res, 2011)21)95-101.

[43] Foolad M,R, Subbiah,P. Kramer,C. Hargrave G, Lin G,Y,:Genetic relationships among cold, salt and drought tolerance during seed germination in an interspecific cross of tomato Euphytica 130 Volume 130, Number 2 (2003), 199-206.

[44] Foolad M,R, Hyman J, R, Lin G, Y. Relationships between cold- and salt-tolerance during seed germination in tomato: Analysis of response and correlated response to selection Plant Breeding. 1999; 118(1) 49–52.

[45] Galaun G,A, Jakobsen K,S, Hughes D,W, The controls of late dicot embryogenesis and early germination. Physiol Plantarum,1991;(81)280-288.

[46] Galloway L,F. Maternal effects provide phenotypic adaptation to local environmental conditions. NEW PHYTOLOGIST,2005;166(1)93-99.

[47] Galloway, Laura F; Burgess, Kevin S.. Artificial selection on flowering time: influence on reproductive phenology across natural light environments. Journal of ecology2012;100(4) 852-861.

[48] Gotwaldová P, Bláha L.: Germinability of minor fodder crops with different provenance under different stress conditions. In: Aktuální poznatky v pěstování, šlechtění, ochraně rostlin a zpracování produktů. Vědecká příloha časopisu Úroda, ISSN 0139–6013, 2008; 171 – 174.

[49] Hong-Bo Shao, Qing-Jie Guo, Li-Ye Chu, Xi-Ning Zhao, Zhong-Liang Su, Ya-Chen Hud, Jiang-Feng Chengc. Colloids and Surfaces B: Biointerfaces Vol. 54, Issue 1, 15 January 2007; 37–45.

[50] Hong-Bo Shao, Qing-Jie Guo, Li-Ye Chu, Xi-Ning Zhao, Zhong-Liang Su, Ya-Chen Hud, Jiang-Feng ChengcColloids and Surfaces B: Biointerfaces 2007; 54; 15(1) 37–45.

[51] Huang W, Ma X, Wang Q, Gao Y, Xue Y, Niu X, Yu G, Liu Y. Significant improvement of stress tolerance in tobacco plants by overexpressing a stress-responsive aldehyde dehydrogenase gene from maize (Zea mays). Plant Mol Biol, 2008;(68) 451-463.

[52] Hundertmark M, Buitink J, Leprince O, Hincha D.K. The reduction of seed-specific dehydrins reduces seed longevity in Arabidopsis thaliana, Seed Sci Res, 2011;(2)165-173.

[53] Hong-Bo Shao, Qing-Jie Guo, Li-Ye Chu, Xi-Ning Zhao, Zhong-Liang Su, Ya-Chen Hud, Jiang-Feng ChengcColloids and Surfaces B: Biointerfaces Vol. 54, Issue 1, 15 January 2007, 37–45.

[54] ISF Value of the Domestic Market for Seed in Selected Countries. On-line. http://www.worldseed.org/isf/seed_statistics.html Accessed 2011.

[55] Karban, R. (2008), Plant behaviour and communication. Ecology Letters,(11) 727–739.

[56] Kovác L. Information and knowledge in biology. Plant Signal Behav. 2007;(2) 65–73.

[57] Karssen C,M. Hormonal regulation of seed development, dormancy, and germination studied by genetic control. In: Seed Development and Germination, J. Kigel adn G. Galili, eds. (New York, Marcel Dekker), 1995;333-350.

[58] Kermode A.R. Regulatory mechanisms in the transition from seed development to germination: interactions beween the embryo and the seed environment. In: Galili, G, Kigel, J. (eds.) Seed development and Germination. New York,1995; 273 – 332.

[59] Kermode A,R, Bewley J,D. Alteration of genetically regulated syntheses in developing seeds by desiccation. In: Leopold, A.C. (ed.) Membranes, Metabolism and Dry Organisms. Cornel University press, New York,1986; 59-84.

[60] Kermode A, R, Finch-Savage B, E. Desiccation sensitivity in relation to seed development. In: Black, M., Pritchard, H.W. (eds.) Desiccation and survival in Plants. CABI publishing, Walingford, 2002; 149-184.

[61] Kosova K; Vitamvas P; Prasil I.T. he role of dehydrins in plant response to cold. Biol Plantarum, 2007;(51):601-617.

[62] Kuchtova P, Vašák J.. The effect of rapeseed stand density on the formation of generative organs. Plant Soil Environ. 2004; 5078–83.

[63] Kvaček Z. et al. Základy systematické paleontologieI. Učební texty, Karolinum, Karlova univerzita, Praha, 2000.

[64] Lester W, Young, Ron W. Wilen; Peta C. Bonham- Smith High temperature stress of Brassica napus during flowering reduces micro- and megagametophyte fertility, induces fruit abortion, and disrupts seed production J. Exp. Bot. 2004;55( 396) 485-495.

[65] Linkies A, Graeber1 K, Knight Ch, Leubner-Metzger G.The evolution of seedsNew Phytologist 2010; (186) 817–831.

[66] Linkies, A., Leubner-Metzger, G. Beyond gibberelins and abscisic acid: how ethylene and jasmonates control seed germination. Plant Cell Report,2012;(31)253-270.

[67] Loıc R, Karine G, Isabelle D, Vandekerckhove J,I, Job C, Job D. The Effect of a-Amanitin on the Arabidopsis Seed Proteome Highlights the Distinct Roles of Stored and Neosynthesized mRNAs during Germination. Plant Physiology,2004;(134)1598–1613.

[68] Metz J., Liancourt P, Kigel J, Harel,D., Sternberg M, Tielbörger K. Plant survival in relation to seed size along environmental gradients: a long-term study from semi-arid and Mediterranean annual plant communities. Journal of Ecology,2010;(98) 697–704.

[69] McClintock B. Significance of responses of the genome to challenge. Science. 1984 Nov 16; 226 (4676) 792-801.

[70] Mokhamed A, M, Raldugina G, N, Kholodova V, P, Kuznetsov V,V. Osmolyte Accumulation in Different Rape Genotypes under Sodium Chloride Salinity. Rus. J. Plant Phys. 2006;(53) 649-655.

[71] Megazyme 2004a. Total Starch Assay Procedure (Amyloglucosidase/α-amylase Method). Firm material Megazyme International Ireland Ltd., Bray, Ireland,.

[72] Megazyme-2004b. Starch Damage Assay Procedure. Firm material Megazyme International Ireland Ltd., Bray, Ireland.

[73] Myers P,N, Setter T,L, Madison J, T., Thompson JF. Endosperm cell division in maize kernels cultured at free levels of water potential. Plant Physiol.1992 99:1051-1056.

[74] Nathan M. Small RNAs: How Seeds Remember To Obey Their Mother CURRENT BIOLOGY Volume: 19 Issue: 15 Pages: R649-R651 Published: AUG (11) 2009.

[75] Nadeau C,D, Ozga J,A, Kurepin L,V, Jin A, Pharis R, P, Reinecke D, M. Tissue-Specific Regulation of Gibberellin Biosynthesis in Developing Pea Seeds. Plant Physiol. 2011;(156)897-912.

[76] Pazderu K., Hosnedl, V. Seed Vigour as Basic Information about Seed Quality. In: Seed and Seedlings: Proceedings from 10th technical and scientific seminar. CULS Prague. 2011;44-48.

[77] Piva G, Bouniols A, Mondiès M.Effect of cultural conditions on yield, oil content and fatty acid composition of sunflower kernel. Proceed. 15th International Sunflower Conference 1. 2000; 62-66.

[78] Schopfer P, Plachy C. Control of Seed Germination by Abscisic Acid 1III. Effect on Embryo Growth Potential (Minimum Turgor Pressure) and Growth Coefficient (Cell Wall Extensibility) in Brassica napus L. Plant physil 1985;77(3) 676-686.

[79] Rikiishi K, Maekawa M, Characterization of a novel wheat (Triticum aestivum L.) mutant with reduced seed dormancy. J Cereal Sci,2010;(51)292-298.

[80] Shabala S. Rhytms in Plants. Springer-Verlag; 2007.

[81] Schoonheim PJ, Pereira D,C, De Boer A.H. Dual role for 14-3-3 proteins and ABF transcription factors in gibberellic acid and abscisic acid signalling in barley (Hordeum vulgare) aleurone cells. Plant Cell Environ, 2009;(32) 439-447.

[82] Scott P. Physiology and Behaviour of Plants. John Wiley & Sons Ltd;2008.

[83] Sun F, Zhang W, Hu H, Li B, Wang Y, Zhao Y, Li K, Liu M, Li X Plant Physiol 2008;(146) 178-188.

[84] Tolleter D, Hincha D,K, Macherel D.A mitochondrial late embryogenesis abundant protein stabilizes model membranes in the dry state. Biochimica et Biophysica Acta–Biomembranes. 2010 (1798)1926-1933.

[85] Toppi L,S.,Skowrońska B,P (EDs). Abiotic stresses in Plants, Kluwer 2003

[86] [86Torices R, Méndez M. Fruit size decline from the margin to the center of capitula is the result of resource competition and architectural constraints. Oecologica, 2010; (164)949-958.

[87] Torices R., Gómez J. M., Méndez M. Where do monomorphic sexual systems fit in the evolution of dioecy? Insights from the largest family of Angiosperms. New Phytologist. 2011; (190) 238-248.

[88] Van Zandt P,A, Mopper, S. The effects of maternal salinity and seed environment on germination and growth in Iris hexagona. Evol Ecol Res 2004;(6)813-832.

[89] Welbaum G.E, Bradford K,J. Water relations of seed development and germination in muskmelon (Cucumis melo L.) I. Water relations of seed and fruit development. Plant Physiol, 1988;(86):406-411.

[90] Young L,W, Ron W, Wilen R,W, Bonham-Smith P,C.High temperature stress of Brassica napus during flowering reduces micro- and megagametophyte fertility, induces fruit abortion, and disrupts seed production J. Exp. Bot. 2004;55(396) 485-495.

[91] Zhang WS (2008) Plant Signal Behav 2008(3): 361-353

Chapter 6

# Abiotic Stress in Plants

Yin Gong, Liqun Rao and Diqiu Yu

Additional information is available at the end of the chapter

http://dx.doi.org/10.5772/55865

## 1. Introduction

Living on the same planet, plants means a lot to us. No matter being taken, as our food or treated with great commercial significance, plants are so indispensable that we have to learn how to protect, make use of, and most important of all, get on well with them. In the first place, what we all understand is: plants are distinguished from us or other animals by being unable to escape from the surrounding circumstances. Thus, when they are confronted with living-threaten pressures, their only choice is to try their best to adjust to them.

The second is "But how?" Plants have developed plenty of physical and biochemical strategies to face up to adverse conditions. Fortunately, thanks to so many excellent researchers' efforts in this field, we have been making so many progresses in identifying and characterizing the mechanisms on how plants perceive outside stress and response to it. Unfortunately, that's far from enough. In this chapter, we will mainly discuss abiotic stress and endeavor to elucidate the mechanism of various reactions plants take at the molecular level.

Before we start our discussion, we probably need to know what abiotic stress is all about. Basically, it includes all the non-living environmental factors that can negatively or even harmfully affect the growth and productivity of plants. Commonly, we choose to put drought, flooding or submergence, salinity, extreme temperatures on our daily researching agenda due to their key roles in producing yield loss of agricultural or industrial crops worldwide. But other kind of abiotic stress is entitled to be paid more attention, such as high light, deficits of inorganic nutrients (nitrogen, phosphorus, potassium et al.), and for sure, they are of definite importance for plants' growth and development. Moreover, one factor we can not set aside is human behavior, which in a large sense put considerable pressure on plants. Residuals of chemicals brought by agricultural practice to improve yield may generate stress, and the increased modification of the atmosphere by human activities is gaining weight. And what we should really stress on is the compounding damaging effects by multiple stress factors acting simultaneously. Thus for plants exposed in diverse stress conditions, struggling for surviving, they organically adapt a complicated interplay of signaling cascade to percepting stress signal, then amplifying, transmitting, and finally triggering stress responses. Furthermore, there do exist overlap between different kinds of stress responses, which truly explains the cross-tolerance phenomenon, a measure taken by plants facing with combining stresses. Here, we are going to introduce signal transduction mechanisms in plants under stressful circumstances, hoping to give readers a general idea about how plants survive in different stressful situations.

## 2. Signal transduction

In general, for plant cells, signal transduction starts from the receptor activation, then the generation of second messengers translating the primary external signal to intracellular signals. These intracellular messengers will be further interpreted by their co-workers resulting in the inspiration of downstream pathways. During the whole process, reversible protein phosphorylation frequently happens; this can activate various transcription factors inducing the expression of stress responsive genes. Moreover, other components are also essential for the pathway to process. They have always been mentioned as signaling partners, mainly working in recruiting and assembling signal complexes, targeting signaling molecules, as well as controlling their lifespan.

Simply saying, the signal transduction pathway is a delicate cooperating process conducted by each single participator including receptors/sensors, second messengers, phosphoprotein cascades, transcription factors, and stress-responsive genes. Eventually the precise and optimal response will be triggered to protect plants from damages in a large sense. As far as we know, signal transduction is indispensable for many cellular activities and their coordination, and most of its steps are complicated occurring in a time and space-dependent manner. In the following part, we are going to explore every single step of signal transduction in order to understand how plants cope with various stress in their lifespan.

### 2.1. Sensors

#### *2.1.1. Complexity in researching on sensors*

Sensors act as the molecules pioneering in perceiving stress stimulus and relaying the signal to downstream molecules to initiate the signal transduction pathway. As the first participator in the pathway, they must be of great researching meaning. However, they are also the mainly intricate role for us to recognize.

Firstly, most of the abiotic stress signals themselves are complicated which probably comprise several physical or chemical signals. Taken the cold stress as an example, it can induce both osmotic stress and mechanistic stress. Similarly, drought may be accompanied by osmotic stress, ionic stress, a mechanistic signal, and heat stress in some cases. Based on these facts, it is natural for us to deduce that for plants there probably exists inequality in treating each single stimulus in accordance with the plant status or the stress severity. That is to say, a simple stimulus may diversely deliver complicated information to the plants, that's exactly why it is very likely for plants to have multiple cellular sensors to perceive each stress signal or one attribute of that signal. Secondly, the redundancy of signal perception make it even harder to identify and confirm each sensor relating to each stress stimulus, since knocking out one receptor may not significantly affect stress signaling outputs. Thirdly, even if we find a putative sensor, how to prove our hypothesis could also become a headache. Because different sensors probably vary in molecular identities, signal-perceiving modes, outputs, and also subcellular localizations, no wonder not much is known about plant abiotic stress sensors.

### *2.1.2. Putative sensors for perceiving stress signal*

First of all, how can an external signal turn out to be internal? Where are the receptors/sensors and transporters? These will be the first bunch of questions we are going to ask. Imaging if we are plant cells, what will be the first weapon we use to maintain inner homeostasis when suffering from the outside disturbances? The answer will probably be "plasma membrane". So far, many researchers have demonstrated that the plasma membrane (PM) is responsible for perceiving and transmitting external stress signals, as well as responding to them. For example, when plants are under salinity stress, salt reaches the PM first, which makes the membrane lipids and transport proteins start to regulate permeability of this membrane triggering primary responses (Cooke and Burden, 1991). In many plants, changes in PM lipids, such as sterols and fatty acids, have been observed responding to salt stress and may contribute to the control of membrane fluidity and permeability, as a primary stress-responsive reaction (Elkahoui et al., 2004). Therefore, it is suggested that physical properties of membranes (lipid composition, fatty acid composition) may lead us to find potential sensors perceiving stress signals.

Secondly, let's stress a little bit more on the most common stress signals, cold, drought, and salinity. All of these three stresses have been detected to induce transient $Ca^{2+}$ influx into the cell cytoplasm (Sanders D et al., 1999; Knight, 2000). Thus we can hypothesis channels responsible for this $Ca^{2+}$ influx possibly acting as a sensor for these stress signals. Based on what we have discussed above, signaling reception may involve changes in membrane fluidity and cytoskeleton reorganization, which are also confirmed in early cold signaling (Sangwan et al., 2001; Wang and Nick, 2001). Coincidentally, cold-induced $Ca^{2+}$ influx in plants occurs only after the occurrence of a rapid temperature drop (Plieth et al., 1999). Taken together, physical alterations in cellular structures may activate certain $Ca^{2+}$ channels under cold stress, which indirectly suggests that $Ca^{2+}$ channels might be a putative sensor.

Except the ion channel as a whole, other types of functional proteins can hardly be ignored on the list of sensors. So far, studies on plants and other systems have also identified several kinds of sensors. And in order to find sensors effectively in plant abiotic pathways, we need to borrow the experience and results of researches on other species. It is known that for plants, cold, drought, and salt stresses will all induce the accumulation of compatible osmolytes and antioxidants (Hasegawa et al., 2000). In yeast and in animals, mitogen-activated protein kinase (MAPK) pathways are responsible for producing osmolytes and antioxidants, which are activated by receptors/sensors such as protein tyrosine kinases, G-protein coupled receptors, and two-component histidine kinases. For plants, only histidine kinases may have been explored and clarified in a deeper sense compared with others.

Retrospecting the history of histidine kinase, one important discovery is the cyanobacterium histidine kinase Hik33 (Suzuki et al., 2000) and the *Bacillus subtilis* histidine kinase DesK (Aguilar et al., 2001) being identified as thermosensors. Unfortunately, even if several putative two-component histidine kinases have been found in *Arabidopsis thaliana* (Urao et al., 2000), none of them can be confirmed as thermosensors. However in yeast, a two-component histidine kinase named SLN1 has been identified as a type of membrane protein sensor for osmotic stress perception (Maeda et al., 1994, 1995). And then later researches found out AtHK1, an Arabidopsis histidine kinase, can complement mutations of *SLN1*. Therefore AtHK1 may participate in osmotic stress signal transduction in plants (Urao et al., 1999). In conclusion, understanding the function of putative histidine kinases and their relationship with MAPK pathways not only help us dig out more sensors but also, even more important, find out how they work in signal transduction pathways.

Thirdly, in plants, the receptor-like kinases and G-protein are worthy to be mentioned in the searching for stress signal sensors. Why? The stress hormone abscisic acid (ABA) makes us study on them who may contain putative stress sensors. It is well-known that the ABA is of great significance in stress signaling, thus, to understand how ABA is perceived certainly will contribute to revealing the hidden sensing-processes of stress signals. Generally, the researches on ABA perception mechanisms always relate to putative receptor-linked components or those putative receptor molecules regulated by stress or ABA.

Here we are going to mention a different way of osmolyte production that involves the pathways triggering the activation of late embryogenesis-abundant (*LEA*)-type genes representing damage repair processes (Zhu, 2001; Xiong and Zhu, 2002). And these LEA-like genes under cold, drought, and salt stress are modulated by phosphoinositols who are closely connected with the activity of phospholipase C, which in plants might be regulated by G-proteins. Moreover evidences suggest G-protein coupled receptors may take part in perceiving a secondary signal derived from these stresses (Ullah et al., 2001; Wang et al., 2001), which may brings a hint that G-protein may have a position on primary sensors list.

On the other hand, *Arabidopsis* heterotrimeric G-protein$\alpha$subunit GPA1 may be a part of ABA response in guard cells but has no relation with ABA-induced stomata closure. Moreover GPA1 interacts with the G-protein couple receptor-like protein GCR1. Researches on gpa1 mutants and gcr1 mutants bring us much more information on finding receptors for ABA. Also some small G proteins are referred to as negative regulator of ABA responses in Arabidopsis, like ROP10. But the really surprising discovery comes into the world in 2009. In that year, two research groups from the USA and Germany reported in Science that they had identified a small protein family binding to ABA interacts with ABA Insensitive 1 and 2 (ABI1 and ABI2), two type 2C protein phosphatases (PP2Cs). And they are negative regulators of ABA signaling (Ma et al. 2009, Park et al. 2009).

With the proceeding of the pathway, we can see that abiotic stresses also give birth to second signaling molecules (discussed below). Therefore, in the next part, we are going to pay attention to the second messengers and their performance in signal transduction pathways.

## 2.2. Second messengers

Several second messengers are active participators in stress signal transduction. Mainly they are groups of small intracellular signaling molecules or ions, normally locating in the cytoplasm of a cell and responding to a signal received by a cell-surface sensor, which activates various kinases to regulate other enzymes' activities. What we mention frequently as second messengers are reactive oxygen species, lipid phosphates-derived signals, and cyclic nucleotides-related signals. Besides, some plant hormones also work as secondary signal molecules under stress conditions.

### *2.2.1. Reactive oxygen species (ROS)*

ROS are species of oxygen which are in a more reactive state than molecular oxygen, resulting from excitation or incomplete reduction of molecular oxygen. Generally, ROS contains both free radical ($O_2$•−, RO•, $HO_2$•, OH•), and non-radical forms ($H_2O_2$, $^1O_2$). For plants, they tend to be a two-edged weapon. On one hand, they are highly reactive and toxic, always taken as unwelcome harmful by-products of normal cellular metabolism, and causes damage to proteins, lipids, carbohydrates, DNA which ultimately results in cell death in plants. On the other hand, it has also been proved that ROS can affect genes' expression and signal transduction pathways, which mean that cells may use it as biological stimuli and signals to activate and regulate various genetic stress-response processes (Foyer and Noctor 2009).

Since it means a lot to plants' life, where and how it can be produced? In photosynthetic tissues, the chloroplast is the prime source of ROS. But for the non-photosynthetic tissues, mitochondria are the leader in production. In chloroplasts, photosystem I and II (PSI and PSII) are the major sites for the production of $^1O_2$ and $O_2$•−. In mitochondria, complex I, ubiquinone and complex III of electron transport chain (ETC) are the major sites for the generation of $O_2$•−. In addition to the mitochondria and NADPH oxidases, additional cellular sources of ROS production include a host of other intracellular enzymes such as xanthine oxidase, cyclo-oxygenases, cytochrome p450 enzymes, and lip-oxygenases for which oxidants act as part of their normal enzymatic function.

Consequently, when plant cells are under stresses, the rate of ROS production usually goes up, inspiring the activities of antioxidants and scavenging enzymes to keep plants live a healthy life. Fortunately, plant cells possess very efficient enzymatic (superoxide dismutase, SOD; catalase, CAT; ascorbate peroxidase, APX; glutathione reductase, GR; monodehydroascorbate reductase, MDHAR; dehydroascorbate reductase, DHAR; glutathione peroxidase, GPX; guaicol peroxidase, GOPX and glutathione-S- transferase, GST) and non-enzymatic (ascorbic acid, ASH; praline; glutathione, GSH; phenolic compounds, alkaloids, non-protein amino acids and a-tocopherols) antioxidant defense systems cooperatively working on controlling the cascades of uncontrolled oxidation and protecting plant cells from oxidative damage by scavenging of ROS. Eventually, the equilibrium has maintained between ROS production and antioxidant defense systems.

However, this balance will always be perturbed by various biotic and abiotic stress factors such as salinity, UV radiation, drought, heavy metals, temperature extremes, nutrient deficiency, air pollution, herbicides and pathogen attacks. Once it has been challenged, various signals pathways start to be proceeded to mediate the disturbances to protect cells from harm brought by extra ROS. For example, when osmotic stress comes, various plant species show an obviously reduced assimilation rate due to stomatal closure (Huchzermeyer and Koyro 2005). This result can be owed to an excessive production of reactive oxygen species (ROS) who are highly destructive to lipids, nucleic acids, and proteins (Kant et al. 2006; Türkan and Demiral 2009; Geissler et al. 2010).

First and foremost, having been identified as second messengers how does ROS affect stress signal transduction? Several enzymes which are involved in cell signaling mechanisms are also potential targets of ROS. These include guanylyl cyclase (E. Vranova, S. Atichartpongkul, 2002), phospholipase C (C.H. Foyer, G. Noctor, 2003), phospholipase A2 (I.M. Moller, 2001) and phospholipase D (A.G. Rasmusson, K.L. Soole, 2004). Ion channels may be targets as well (G. Noctor, R.D. Paepe, 2006), among which calcium channels was mentioned (D.M. Rhoads, A.L. Umbach, 2006). Since calcium has ubiquitous functions in plant stress signal transduction pathway, we are interested in the relationship between ROS and calcium. Before dive into calcium, let's back to NADPH oxidases that are an important ROS-generating system. RBOHs shorting for respiratory burst oxidase homologs is always an eye-catching topic. Recent evidence points out RBOHs relate to heavy-metal induced accumulation of ROS (Pourrut et al. 2008) and early response to salt stress (Leshem et al. 2007). Subsequently, ROS produced by Rbohs are thought to activate $Ca^{2+}$ channels leading to further increases in cytosolic $Ca^{2+}$ (Foreman et al. 2003) and downstream signaling. In general, it has been suggested that ROS took part in the regeneration of $Ca^{2+}$ signals by activating $Ca^{2+}$ channels. Then additional signal transduction was triggered through $Ca^{2+}$-mediated pathways (reviewed in Mori, I.C. and Schroeder, J.I. 2004).

Except the interaction with calcium, another route for ROS to work is that ROS themselves can directly modify signaling molecules through redox regulation. Redox status inside a cell is essential to the correct functioning of many enzymes, which can be used to alter enzyme activity; thus alteration of the redox status could be treated as a signaling mechanism (Gamaley and Klyubin, 1999). One of the most important and well-known redox-sensitive molecules in this respect is glutathione (GSH), which can form the GSH/GSSG couple. The balance between the GSH and GSSG takes the central position to maintain cellular redox state (C.H. Foyer, G. Noctor, 2005). But ROS like H2O2 can affect the process of lowering the cells' GSH content to alter the redox status. Meanwhile, it also has been suggested that enzymes such as ribonucleotide reductase and thioredoxin reductase, as well as transcription factors, might be among the targets for altered redox status. In detail, cysteine residues of molecules may be key active sites as targets for redox regulation. And these molecules can act as potential sensors for ROS (Xiong, L. and Zhu, J.K. 2002).

Secondly, ROS are very likely to play a significant role in the activation of stress-responsive genes, especially those who encode enzymes responsible for antioxidants biosynthesis or enzymes directly detoxify reactive oxidative radicals. For example, $H_2O_2$ production is thought to be raised under various abiotic stresses, which can enhance gene expression of active oxygen scavenging (AOS) enzymes. NO, produced under salt stress, could serve as a second messenger for the induction of PM H-ATPase genes' expression, which promote PM H-ATPase activity (Liqun Zhao, Feng Zhang et al., 2004).

Thirdly, we are going to stress a little more on $H_2O_2$ and NO. In maize, $H_2O_2$ production grows up induced by chilling stress, and exogenously applied $H_2O_2$ lifted up chilling tolerance (T.K. Prasad, M.D. Anderson, 1994). Increased $H_2O_2$ production has been detected occurring gradually responding to salt stress in rice plants (N.M. Fadzilla, R.P. Finch, 1997). Moreover, $H_2O_2$ was also reported to induce small heat shock proteins (HSP26) in tomato and rice (J. Liu, M. Shono, 1999; B.H. Lee, S.H. Won, H.S. Lee, 2000). However, it was recently shown that $H_2O_2$ produced by apoplastic polyamine oxidase can influence the salinity stress signaling in tobacco and can play a role in balancing the plant response between stress tolerance and cell death (Moschou et al. 2008). NO has also been suggested to act as a signal molecular mediating responses to biotic and abiotic stresses. Under salt stress, NO could serve as a second messenger for the induction of PM H-ATPase expression, which may account for the enhanced PM H-ATPase activity. Thus, ion homeostasis is reestablished so as to adapt to salt stress (Liqun Zhao, Feng Zhang et al., 2004).

Furthermore, researches on ABA give us much more information on $H_2O_2$ and NO in signal transduction. So let's take a look at how they work in ABA signaling and other signal transduction pathways. The process of stomata closure regulated by ABA in a large sense require the generation of $H_2O_2$. Moreover, $H_2O_2$ production may be a prerequisite for ABA-induced stomatal closure (Zhang, X., Zhang, L. et al. 2001). Experiments have found out mutations in genes encoding catalytic subunits of NADPH oxidase, known as the major source for $H_2O_2$ production, will impair ABA-induced ROS production, as well as the activation of guard cell $Ca^{2+}$ channels and stomata closure (Kwak J.M., Mori, I.C. et al. 2003). In plants, both nitrate reductases and NO synthases (NOS) can contribute to NO generation. Loss-of-function mutations in Arabidopsis NOS, AtNOS1, impair ABA-induced NO production and stomata closure (Guo, F.Q., Okamoto, M. and Crawford, N.M., 2003).

On the other hand, accumulated evidence indicate ROS seem to play a central role in regulating Mitogen-activated protein kinase (MAPK or MPK) cascades (discussed later in detail). However, only the functions of MPK4, MPK3 and MPK6, out of the 20 Arabidopsis MAPKs, have been thoroughly characterized. What really counts is they can all be activated by ROS and abiotic stress. In addition, activities of MPK1 and MPK2 have been shown to be provoked by $H_2O_2$ and ABA (Ortiz-Masia et al. 2007). Also MPK7 was found to be activated by $H_2O_2$ under specific circumstances (Dóczi et al. 2007). Furthermore, the authors found that $H_2O_2$ may probably have a generally stabilizing impact on MAPKKs (MAP kinase kinases). In the future, we still have lots of work to elucidate how ROS regulate MAPK signaling in abiotic stress field.

All in all, even if we are holding a lot of evidence about the functions of ROS in abiotic stress, we still have to face up to those obscure steps relating to different mechanisms, not to mention hundreds of stress responsive genes involved in.

### *2.2.2. Lipid-derived signal messengers*

It is well-known that cellular membranes contains a wide range of different lipids, including sphingo-, neutral-, glyco-, and phospholipids, all with unique biophysical properties. Beyond the structural role, some of them are equipped with direct signal-transducing properties. What we discussed in the sensors part is that membrane lipids can directly response to abiotic stress stimuli by modulating membrane fluidity or its other physiochemical properties, but in this part we will take another angle to demonstrate its significant function in the process of generating intracellular signaling molecules. Moreover lipids and their biogenesis and degradation enzymes play many direct or indirect roles to regulate or affect signaling and stress tolerance. In signal transduction, signaling lipids are distinguished for their low abundance and rapid turnover. They are rapidly formed responding to diverse stimuli through lipid kinases or phospholipases' activation. Thanks to the lipid-binding domains, these lipid signals can activate enzymes or recruit proteins to membranes leading to the activation of downstream signaling pathways resulting in specific cellular events and physiological responses. Studies on them find out, for plants, lipid signaling form a complex regulatory network responding to abiotic stress.

Basically, in eukaryotes, typical signaling lipids includes phosphatidylinositol lipids (polyphosphoinositides; PPIs), certain lyso-phospholipids, diacylglycerol (DAG), and phosphatidic acid (PA) (Munnik and Testerink, 2009; Xue et al.,2009; Munnik and Vermeer, 2010). Among them, PA is of great importance as a lipid second messenger in plants involved in various biotic and abiotic stress conditions. Based on thousands of researches, it is easy to detect almost every environmental cue can trigger a rapid PA response (Testerink and Munnik, 2005; Arisz et al., 2009; Li et al., 2009; Mishkind et al., 2009). How can PA be produced? Two ways in brief. Directly PA is generated through activation of phospholipase D (PLD), and indirectly a phospholipase C/diacylglycerol kinase (PLC/DGK) pathway regulated by two types of PLC enzyme named as the PI-PLCs (phosphoinositide-PLCs) and NPCs (non-specific PLCs). After the rapidly bounce up under stress, PA level will go back to normal when stimuli disappear. (Christa Testerink et al. 2011).

In most of the osmotic stress cases, both PLC/DGK and PLD pathways are activated leading to fast and transient PA accumulation, but exceptions also exist (Zonia and Munnik, 2004; Darwish et al., 2009; Hong et al., 2010). Besides responding to osmotic stress, we also see the PLD$\alpha$1 enzyme participating in cold, frost, and wound stress signaling (Bargmann et al., 2009; Hong et al., 2010) and probably by promoting responses to ABA, especially in stomata (Mishra et al., 2006). On the other side, PLC/DGK pathways also get activated by salinity (Arisz, 2010). Earlier we knew that AtPLC1, one of the PI-PLCs, was shown to be induced by salinity and drought (Hirayama et al., 1995), which is necessary for ABA-induced inhibition of germination and gene expression (Sanchez and Chua, 2001). Recently, NPC4 (a NPC isoform) was found to modulate responses to ABA and bring enhanced salt and drought tolerance (Peters et al., 2010). In addition, several ABA signaling proteins have been identified as potential PA targets (Mishra et al., 2006), which has further suggested PA could mediate ABA responses. Meanwhile, cooperation between the NADPH oxidase isoforms RbohD, RbohF and PA brings more information for us to understand PA's function in ABA-induced ROS generation and stomatal closure (Zhang et al., 2009). Furthermore, PA also targets other protein kinases like SnRK2 protein kinase (Testerink et al., 2004), MAPK isoform MPK6 (Yu et al., 2010), sphingosine kinase (SPHK) (Liang Guo, XueminWang, 2012) to influence diverse signaling transduction pathways.

Other two important second messenger molecules - inositol phosphates Ins(1,4,5)P3 and DAG (diacylglycerol) are worthy to be mentioned here. Firstly, InsP3 was shown to release $Ca^{2+}$ from an intracellular store in the early 90s, but recently evidences pops up suggesting that InsP6 was shown to release $Ca^{2+}$ at a 10-fold lower concentration than InsP3 (Lemtiri-Chlieh et al., 2003; Teun Munnik, 2009), who can be generated by phosphorylating InsP3. By the way data can be found demonstrating that whole-plant IP3 level goes up significantly within 1 min after stimuli occur, and keep the tendency for more than 30 min under stress. Under osmotic stress, in Arabidopsis, phosphatidylinositol 4,5-bisphosphate (PtdIns(4,5)$P_2$) is hydrolyzed to IP3. And IP3 accumulation occurs coincidently in a time frame similar to stress-induced calcium mobilization (Daryll B. DeWald et al., 2001). In conclusion, what we can say is under different stress, to identify the most critical second messenger molecules depends on the research on the network consisting of multiple polyphosphoinositides.

Secondly, diacyglycerol (DAG) is an important class of cellular lipid messengers, but for its function in plants, data is not sufficiently provided. In *Arabidopsis thaliana,* knocking out NPC4 results in DAG level decrease and compromises plant response to ABA and hyperosmotic stresses. On the other hand, overexpressing NPC4 leads to higher sensitivity to ABA and stronger tolerance to hyperosmotic stress than wild-type. And later experiments indicate that NPC4-produced DAG is converted to PA and NPC4 might be a positive regulator in ABA response and promote plant tolerance to drought and salt stresses (Carlotta Peters et al, 2010). Furthermore, all higher plant genomes sequenced so far lack both InsP3 receptor and the DAG target, PKC (Munnik & Testerink 2009). In conclusion, we have reasons to believe in that PA rather than DAG are more likely to play a central role in stress signaling transduction.

At last, we'd love to say more about other types of phospholipases like secreted phospholipase $A_2$ and patatin-related phospholipase A (pPLA) who were shown to have functions in auxin signal transduction by cooperating with auxin receptors ABP1 or TIR1. (Günther F. E. Scherer et al., 2012 ). And this fact helps us to further confirm the significance of phytohormones in signaling transduction which will be discussed below.

### 2.2.3. *Phytohormones*

When it comes to phytohormones, strictly speaking, we can not conclude them as second messengers, but the first and foremost idea needs to be posted here is "The most powerful players in intercellular regulation are plant hormones." (Wolfgang Busch, Philip N. Benfey, 2010). Owing to the broad and diverse functions, many of phytohormones were discovered before the dawn of molecular genetics (Sachs and Thimann, 1967; Thimann and Skoog, 1933). In generation, they are a large bunch of trace amount growth regulators, the best-known group comprises auxin (IAA), cytokinin (CK), gibberellic acid (GA), abscisic acid (ABA), jasmonic acid (JA), ethylene (ET), salicylic acid (SA), but the name list is growing by time. Here we add brassinosteroids (BR), nitric oxide (NO), polyamines, and strigolactone (SL). Indubitably phytohormones have various functions in growth and development. Indeed, they play central roles in nutrient allocation, and source/sink transitions. However based on former description relating to sensors and second messengers, we have to focus on here is that these low-molecular-weight compounds are indispensable for coordinating various signal transduction pathways during responses to various abiotic stresses.

Firstly, they function as systemic signals that can transmit information over large distances. Like ABA, it can be transported and play physiological roles at sites far away from where it is synthesized (Sauter, A. et al., 2001). And different types of cells have their own understanding even for the same hormones signal. And information from diverse hormones always triggers coherent responses of cells. This is signal perception at cellular level. Lucky for us, modern transcriptome profiling technologies have provided a global view of hormones' effects at the molecular level and identified hundreds to thousands of genes, the expression levels of which are modified by individual hormones (Goda et al., 2008). A large number of data have proved that treating plants with exogenous hormones will rapidly and transiently alter genome-wide transcript profiles (Chapman and Estelle 2009).

Secondly, complex networks of gene regulation by phytohormones under abiotic stresses involve various *cis*- or *trans*-acting elements. Some of the transcription factors regulated by phytohormones include ARF, AREB/ABF, DREB, MYC/MYB, NAC, WRKY and other key components functioning in signaling pathways of phytohormones under abiotic stresses will be briefly mentioned later. And they often rapidly alter gene expression by inducing or preventing the degradation of transcriptional regulators via the ubiquitin–proteasome system (Santner A, Estelle M, 2010).

Thirdly, the ability of plants to a wide range of environmental stresses is also finely balanced through the interaction of hormonal plant growth regulators and the redox signaling hub. Plant hormones produce reactive oxygen species (ROS) as second messengers in signaling cascades that convey information concerning changes in hormone concentrations and/or sensitivity to mediate a whole range of adaptive responses (Carlos G. Bartoli et al., 2012). For example, Brassinosteroids (BRs) can induce plant tolerance to diverse abiotic stresses by triggering $H_2O_2$ generation in cucumber leaves (Cui et al., 2011). In the following part we will simply introduce how phytohormones work in signal transduction, and how they talk with each other when they exchange information.

Let's start from ABA, whose synthesis is one of the fastest responses to abiotic stress for plants. Under water stress, ABA synthesis triggers ABA-inducible gene expression leading to stomatal closure, thereby reducing water loss through transpiration, and consequently, a reduced growth rate (Schroeder, J.I. et al., 2001). ABA also plays a vital role in adapting to cold temperatures. Cold stress induces the synthesis of ABA and the exogenous application of ABA enhances the cold tolerance of plants. A large number of genes associated with ABA biosynthesis and ABA receptors-encoding genes and downstream signal relays have been characterized in Arabidopsis thaliana (reviewed by Cutler SR et al., 2010). ABA activates the expression of many stress-responsive genes independently or synergistically with stresses, which makes it become the most studied stress-responsive hormone.

Other hormones, in particular CK, SA, ET, and JA, also have substantial direct or indirect performances in abiotic stress responses. CK is an antagonist to ABA, and under water shortage situation, CK levels usually decrease. But transgenic tomato rootstocks expressing IPT (isopentenyl transferase, a gene encoding a key step in CK biosynthesis) had improved root CK synthesis shown raised salinity stress tolerance (Ghanem ME et al., 2011). Meanwhile, by checking public microarray expression data for A. Thaliana, numerous genes encoding proteins associated with CK signaling pathways have been found affected by various abiotic stresses (Argueso CT et al., 2009). Although auxins, GAs, and CKs have been implicated primarily in developmental processes in plants, they regulate responses to stress or coordinate growth under stress conditions (Günther F. E. Scherer et al., 2012; F. Eyidogan et al., 2012). Auxins taking part in drought tolerance was postulated by researchers (ZhangS-W et al., 2009). What's more, BR was reported (mainly researches on exogenous application of BR) to induce stress-related genes' expression, which results in the maintenance of photosynthesis activity, the activation of antioxidant enzymes, the accumulation of osmoprotectants, and the induction of other hormone responses (Divi UK et al., 2009). In conclusion, there do exist a complex network for phytohormones to contribute to stress-induced reactions for plants. And due to the overlap between hormone-regulated gene suites in the adaptive responses, we have to discuss cross-talk between the different hormone signaling pathways as a extensive part of that complex network.

Earlier, it was reported that ABA can inhibit the biosynthesis of ethylene and may also potentially reduce the sensitivity of plants to ethylene (Sharp, R.E., 2002). Recently the expression of many other genes associated with auxin synthesis, perception, and action has been shown to be regulated by ethylene (Stepanova AN, Alonso JM, 2009). And it is not surprising that auxin has been found involved in ethylene biosynthesis very early. Meanwhile, CK was shown to be a positive regulator of auxin biosynthesis (Jones B et al., 2010).

Furthermore, GA and BR regulate many common physiological processes like the growth and development in rice seedlings (Wang L et al., 2009). Except BR, GA has another partner - SA. Transgenic A. thaliana plants constitutively overexpressing a GA-responsive gene became more tolerant under abiotic stress and this stronger tolerance was correlated with increased endogenous levels of SA (Alonso-Ramirez A et al., 2009).

Discussed above, ABA can regulate stomatal actions under stress conditions; however, it is not alone in that process. CK, ET, BR, JA, SA, and NO also affect stomatal function (reviewed by Acharya B, Assmann S, 2009). In detail, ABA, BR, SA, JA, and NO induce stomatal closure, CK and IAA promote stomatal opening. And we mentioned before, NO acts as a key intermediate in the ABA-mediated signaling network in stomatal closure. Moreover BR-mediated signaling was regulated by ABA, and in turn, ABA was also shown to inhibit BR-induced responses under abiotic stress (Divi U et al., 2010). And it is not hard to deduce that there are other tricky relationships between different hormone-involved pathways. Cross-talk between the phytohormones results in synergetic or antagonist interactions, which is crucial for plants in abiotic stress responses. To characterize the molecular mechanisms regulating hormone synthesis, signaling, and action means a lot to modificate hormone biosynthetic pathways to develop transgenic crop plants with promoted tolerance to abiotic stress.

### 2.3. $Ca^{2+}$ as an intermediate signal molecule

So far, we have already touched some grounds related to Calcium ($Ca^{2+}$) functioning in signal transduction. Among all ions in eukaryotic organisms, it is likely to be the most versatile one who almost links to all aspects of plant development, not to mention many regulatory processes. The reason why it is so powerful may root in its flexibility in exhibiting different coordination numbers and complex geometries, and this ability makes it easily form complexes with proteins, membranes, and organic acids. However we won't detect high cytosolic or organelle $Ca^{2+}$ concentrations resulting from the tight management of various $Ca^{2+}$ pumps and transporters. The reason why the concentration needs to be controlled is that higher $Ca^{2+}$ concentrations can chelate negatively charged molecules in the cell leading to cytotoxicity. Interestingly, all the secondary signaling molecules we mentioned above may activate transient increases in cytosolic $Ca^{2+}$, and transient elevations in cytosolic $Ca^{2+}$ concentration have been documented to have relationship with a multitude of physiological processes linking to abiotic stress responses. So we may wonder concentration control probably will help us tell the story of another famous second messenger--- $Ca^{2+}$ in signal transduction for plants under abiotic stress.

Earlier in 1982, research on the green algae Chara told us the cytosolic $Ca^{2+}$ concentration change predicted $Ca^{2+}$ might work as a second messenger in plants (Williamson and Ashley, 1982). Based on later reports, it has been found various stimuli will spur their own special $Ca^{2+}$ responses differing in where and how changes happen (Johnson et al., 1995; Tracy et al., 2008), which exactly supports the former concept of $Ca^{2+}$ signature. For plants, to maintain $Ca^{2+}$ homeostasis, they need the help from $Ca^{2+}$ channels, pumps, and exchangers (carriers) to make specific adaptation to every kind of stimulus (Kudla et al., 2010). Later, cellular $Ca^{2+}$ signals are decoded and transmitted by $Ca^{2+}$-binding proteins that relay this information into downstream responses. Major $Ca^{2+}$ signal transduction routes contain $Ca^{2+}$-regulated kinases mediating phosphorylation events and regulation of gene expression via $Ca^{2+}$-regulated transcription factors and $Ca^{2+}$-responsive promoter elements.

Generally speaking, $Ca^{2+}$ signaling comprises three phases: generation of a $Ca^{2+}$ signature, sensing the signature and transduction of the signal (Reddy and Reddy, 2004). Having discussed above, we are informed that the concentration change are always triggered by cellular second messengers, such as NAADP, IP3, IP6, Sphingosine-1-Phospate, and cADPR (Allen and Sanders, 1995; Navazio et al., 2000; Lemtiri-Chlieh et al., 2003). Then a specific cellular $Ca^{2+}$ signature is sensed by $Ca^{2+}$-binding proteins, the $Ca^{2+}$ sensors (Dodd et al., 2010; Reddy and Reddy, 2004). The sensors themselves may become active to transduce the signal by themselves, or choose to bind to their interacting proteins and affect their partners' activity to transduce the signal. In detail, there are three major classes of $Ca^{2+}$ sensors identified in plants. The first one is calmodulins (CaMs) and calmodulin-related proteins (CMLs). CaMs is a group of small acidic protein, highly conserved in eukaryotes (Snedden and Fromm, 2001), and contains four EF hands (one major $Ca^{2+}$ binding motif) where bind $Ca^{2+}$. Moreover this binding action induces a conformational change of CaM, leading to exposure of hydrophobic surfaces and further triggering electrostatic interactions with target proteins - CaM-binding proteins (CBPs) (Hoeflich and Ikura, 2002). CBPs have been found to take part in regulating transcription, metabolism, ion transport, protein folding, cytoskeleton-associated functions, protein phosphorylation and dephosphorylation, as well as phospholipid metabolism (Yang and Poovaiah, 2003; Reddy and Reddy, 2004). Furthermore, different CaM proteins exhibit differential expression and are likely to show differential affinity to $Ca^{2+}$ and to their target proteins (McCormack et al., 2005; Popescu et al., 2007), which makes CaM be equipped with multiple capabilities in $Ca^{2+}$ signal transduction. With many similarities to CaMs, CMLs are mostly composed of four EF-hands and lack other known functional domains. Like CaMs, they relay the signal by binding to other proteins resulting in activation or inactivation of interacting proteins. Over 300 proteins that interact with CaMs and CMLs have been identified in plants (Popescu et al., 2007). The second class of $Ca^{2+}$ sensor is represented by the calcium-dependent protein kinases (CDPKs/CPKs) who are serine/threonine protein kinases contain a catalytic kinase domain and EF-hand motifs (Cheng et al., 2002). The third typical sensor type is the EF-hand-containing $Ca^{2+}$-modulated protein named SCaBP (SOS3 (Salt-Overly-Sensitive 3)-like $Ca^{2+}$-biniding proteins)/Calcineurin B-like (CBL) proteins, which is plant-specific (Luan et al., 2002). CBLs interact with a family of protein kinases called CBL-interacting protein kinases (CIPKs) (Luan et al., 2009; Weinl and Kudla, 2009; Batistic et al., 2010). In addition to EF-hand-containing $Ca^{2+}$ binding proteins, there are other proteins without that motif acting as sensors who also can bind $Ca^{2+}$, like PLD (introduced in 2.2.2), annexins and C2 domain–containing proteins (Clark and Roux, 1995; Reddy and Reddy, 2004; Laohavisit and Davies, 2011), however their functions in abiotic stress responses haven't been deeply explored, and only some reports suggest PLD and annexin be relevant to stress signal transduction (White et al., 2002; reviewed by Laohavisit and Davies, 2011).

However here we will mainly stress on the EF-hand-containing sensors due to their considerable significance in signal transduction pathways. Based on their functional styles, these sensors are assigned into two camps termed as sensor relays and sensor responders (Kudla et al., 2010). The sensor relays do not have any known enzymatic or other functional domains except the EF hands. They interact with other proteins and regulate their activities, just like CaMs/CMLs, and CBLs (with one exception, CaM7) (McCormack et al., 2005; Luan, 2009; DeFalco et al., 2010). The members of another camp are characterized by an additional a catalytic or functional domain, except EF hands, whose activity is regulated by $Ca^{2+}$ binding to EF-hand motifs. So definitely CDPKs belong to this camp, and other members are

$Ca^{2+}$-and $Ca^{2+}$/CaM-dependent protein kinases (CCaMKs), some DNA or lipid binding proteins, and a few enzymes (Day et al., 2002; Yang and Poovaiah, 2003; Harper and Harmon, 2005). Many calcium sensors are coded by multiple genes, and expression of many of these is induced by stresses (DeFalco et al., 2010).

Then, let's take a look at what will be the next move of those $Ca^{2+}$-activated sensors. Firstly, they can directly bind to *cis*-elements in the promoters of specific genes and induce or repress their expression. Secondly, they will choose to bind to DNA binding proteins and activate or inactivate them, thereby resulting in activation or repression of gene expression. The third path belongs to the activated $Ca^{2+}$-regulated protein kinases (CDPK,CaM binding protein kinase (CBK), CIPK and CCaMK) or phosphatases. They phosphorylate/ dephosphorylate a transcription factor (TF), respectively, resulting in activation or repression of transcription, which allow for the perception and transmission of $Ca^{2+}$ signatures directly into phosphorylation cascades that orchestrate downstream signaling responses (Weinl and Kudla, 2009). The most well-known TFs involved in the phosphorylation include Calmodulin binding transcription activators (CAMTAs; also referred to as signal-responsive proteins or ethylene-induced CaM binding proteins) (Reddy et al., 2000), MYB family (Popescu et al., 2007), the WRKY family (Park et al., 2005; Popescu et al., 2007), basic leucine zipper (bZIP) TFs, like TGA3 (Szymanski et al., 1996) and ABA-responsive TFs ABF1, 2, 3, and 4 (reviewed in Galon et al., 2010), and CBP60s who is a plant-specific CaM binding proteins family (Reddy et al., 1993; Zhang et al., 2010), as well as members of NAC family (Kim et al., 2007; Yoon et al., 2008). However they are TFs binding to $Ca^{2+}$/CaM under $Ca^{2+}$ regulation, so we posted here another two TFs who can directly bind $Ca^{2+}$. One is encoded by *Arabidopsis NaCL-INDUCED GENE (NIG)* (Kim, J., and Kim, H.Y., 2006), and the other is At-CaM7 (Kushwaha et al., 2008). In sum, $Ca^{2+}$ and their interacting proteins served as the upstream elements play an important role in regulating of some stress genes expression.

Meanwhile, what performance they will give in each specific abiotic stress condition needs to be simply introduced here. For drought stress, cellular $Ca^{2+}$ transmits drought signals to regulate the physiological responses induced by drought stress (Dai et al., 2007). It has been found that $Ca^{2+}$ treatment increased protection against membrane lipid peroxidation and stability of membranes and therefore resulted in the increase of drought resistance of rice seedlings. It is also reported that in wheat $Ca^{2+}$ may reduce the adverse stress effects by elevating the content of proline and glycine betaine, thus improving the water status and growth of seedlings and minimizing the injury to membranes (Geisler et al., 2000; Munns et al., 2006; Goldgur et al., 2007). Additionally, $Ca^{2+}$/CaM means a lot to the process of ABA-induced drought signal transferring under PEG stress. And ABA synthesis correlates with cytoplasmic $Ca^{2+}$ concentrations ($[Ca^{2+}]_{cyt}$) (Rabbani et al., 2003; Noctor, 2006). We know how important ABA is to the stomatal status, and now more studies have established a close relationship between $[Ca^{2+}]_{cyt}$ oscillation and stomatal status. In addition, in Arabidopisis genome, 9 SOS3 homologs (SCaBP/CBL) and 22 SOS2 homologs (SOS2-like protein kinases - PKS/CBL-interacting protein kinases-CIPK) were identified. By the way, SOS2 is a serine/threonine protein kinase with an SNF1/AMPK–like catalytic domain and a unique regulatory domain (Liu et al., 2000). Individual SCaBP/CBL interacts with PKS/CIPK with different specificities (Gong et al., 2004; Luan et al., 2002). And it is indicatied that the interaction between SCaBP5 and PKS3 may interpret $Ca^{2+}$ signatures resulting from ABA or drought stress signals. On the other hand, SOS3 interact with and activate the SOS2, whose mutation also confers salt sensitivity. Then the activated SOS2 phosphorylates and regulates ion transporters such as the $Na^+/H^+$ antiporter SOS1 controlling long-distance $Na^+$ transport from the root to shoot, which eventually leads to the restoration of ion homeostasis in the cytoplasm under salt stress (Zhu, J.K., 2003).

In light of salt stress, like other stresses, it is perceived at cell membrane and then trigger intracellular-signaling cascade including the generation of secondary messenger molecules like $Ca^{2+}$ and protons. For instance, it was found that in barley roots, under NaCl stress, $Ca^{2+}$-CaM system may work in activating tonoplast $H^+$-ATPase and regulating $Na^+$ and $K^+$ uptake with involvement of SOS signal transduction pathway (Brini et al., 2007). In Arabidopsis, experiments to overexpress AtCaMBP25 (a CaM binding protein) who may be a a negative regulator of osmotic stress tolerance find the transgenic Arabidopsis plants show higher sensitivity to osmotic stress, while the antisense plants gain more tolerance under salt stress (Perruc, E. et al., 2004).

Furthermore, when it comes to uncomfortable temperature, is there any position for $Ca^{2+}$? For sure. The $Ca^{2+}$ channels have shown their power for the growth of root hairs and the low temperature acclimation of chilling-resistant plants. That's why we find data indicating that the activity and stability of $Ca^{2+}$-ATPase under 2 °C low temperature are the key factors in the development of cold resistance of winter wheat (Yamaguchi-Shinozaki, 2006). Moreover the studies on Arabidopsis mutants displaying reduced tonoplast $Ca^{2+}/H^+$ antiport (CAX1) activity indicate that CAX1 participates in the development of the cold acclimation response (Lecourieux et al., 2006). On the other field of temperature acclimation, there exists data showing in Arabidopsis, $Ca^{2+}$/CaM have gotten involved in heat shock response (Zhang et al., 2009). And we also find some researches on overexpression of a CDPK in rice which brings increased tolerance to cold and salt stress (Saijo, Y. et al., 2000).

Taken together, depending on the type of signal or the type of cell, internal and/or external $Ca^{2+}$ stores could be involved in raising $[Ca^{2+}]_{cyt}$ (Dodd et al., 2010; Kudla et al., 2010). Both types of $Ca^{2+}$ transporters, namely, $Ca^{2+}$-ATPases and CAXs involve in plant responses by regulating $[Ca^{2+}]_{cyt}$. Based on that, regulating cellular and intercellular $Ca^{2+}$ signaling networks brings improving resistances or tolerances. And seeing from the regulation networks of stress responses to drought, salt and cold stress, we find $Ca^{2+}$ and its interacting proteins may be the cross-talks among ABA-dependent, MAPK and other stress signaling pathways. Anyway, we can see the core of $Ca^{2+}$ actions in relaying abiotic stress signaling depends on how to translate $Ca^{2+}$ signatures to specific protein phosphorylation cascades,which we have mentioned above, so in the following part, we are going to trace the performance of phosphoproteins in signal transduction.

## 2.4. Role of phosphoproteins in stress signaling

By controlling the phosphorylation status of other proteins, protein kinases and phosphatases play a fundamental role in coordinating the activity of many known signal transduction pathways. For many signal pathways not only in abiotic stress field, protein reversible phosphorylation is the major player in relaying signals. And during this significant process, we highlight the functions of protein kinases and protein phosphatases who are enzymes to catalyze these reversible phosphorylation processes. And they are divided into several categories according to their structure or functional characteristics. And in the following part, we will give a general idea to the readers about their central role in signal transduction in abiotic stress aspect.

### *2.4.1. MAPK*

Obviously, we have seen MAPK many times in formal description in this chapter. Even if it is not found in plants, the mitogen activated protein kinase (MAPK) cascades are known to be involved in plant abiotic stress responses acting as intracellular signal modules that mediate signal transduction from the cell surface to the nucleus. The reason to mention it here is that phosphorylation plays a central role in the progression of the signal through the MAPK cascade. Moreover MAPK cascades, the conserved signaling modules found in all eukaryotes, are fundamental in transducing environmental and developmental cues into intracellular responses bringing changes in cellular organization or gene expression. The simplest constitution of a MAPK cascade contains MAP kinase kinase kinases (MAP3Ks/MAPKKKs/MEKKs), MAP kinase kinases (MAP2Ks/MAPKKs/MEKs/MKKs) and MAP kinases (MAPKs/MPKs) (Mishra NS et al., 2006). And when under stress, stimulated plasma membrane will activate MAP3Ks or MAP kinase kinase kinase kinases (MAP4Ks), who may be the adapters to link upstream signaling steps to the core MAPK cascades (Dan I et al., 2001). Following that, MAP3Ks will phosphorylate two amino acids in the S/T-X3-5-S/T motif of the MAP2K activation loop. Then MAP2Ks phosphorylate MAPKs on threonine and tyrosine residues at a conserved T-X-Y motif at the active site. When signals come to MAPKs, further phosphorylation will tag on a wide range of substrates involving other kinases, cytoskeleton-associated proteins, and/or transcription factors. As for formation and integrity of a specific MAPK cascade, scaffold proteins take control over it (Whitmarsh AJ et al., 1998). And after signaling completed, MKPs (MAPK phosphatases) take the responsibility to shut the pathway down. Generally, the whole cascade is regulated by various mechanisms, including not only transcriptional and translational regulation but through post-transcriptional regulation such as protein-protein interactions (Rodriguez MC et al., 2010).

Thanks to traditional genetic and biochemical methods and lots of excellent research efforts, we can conclude that MAP3K/MAP2K/MAPK signaling modules show overlapping roles in controlling diverse cellular functions by forming complex interconnected networks within cells. These include cell division, development, hormone signaling and synthesis, and response to abiotic stress (high and low temperature, drought and high and low osmolarity, wounding, high salinity, UV radiation, ozone, ROS, heavy metals), as well as biotic stress reactions (Jonak C et al., 2002; Xiong L et al., 2003; Raman, M. et al., 2007; Gohar Taj et al., 2010; Alok Krishna Sinha et al., 2011).

According to the researches on MAPK pathways, we can see that regulated expression of MAPK components shows effects on stress sensitivity. Here are some examples. Expression of an active form of a tobacco MAP3K, NPK1, increases freezing tolerance of transgenic tobacco or maize plants (Kovtun, Y et al., 2000; Shou, H et al., 2004). Meanwhile, MAP2K1 shows transcriptional induction under salt stress, drought and cold, as well as activated by wounding and drought stress. And MAP2K1 can phosphorylate MAPK4. An unsurprised fact is that MAPK4 and MAPK6 are found to be activated by cold, salt and drought (Ichimura K et al., 2000). Indeed, a MAPK module composed of MAP3K1-MAP2K1/MAP2K2-MAPK4/MAPK6 has been confirmed in cold and salt stress by yeast two hybrid analyses and yeast complementation (Teige M et al., 2004). So we can say different MAPKs are activated at different times after the onset of stress and the activities of these MAPKs are activated within different time periods. By the way, during osmolarity signaling MAPKs module seems to be widely involved (reviewed by Gohar Taj et al., 2010).

Due to the interlink between osmotic stress and oxidative stress, we are informed of the relationship between ROS, hormone signaling and MAPKs. ROS like $H_2O_2$ is closely associated to MAPKs' activities. In Arabidopsis, $H_2O_2$ activates AtMPK6 and the related AtMPK3 via the MAP3K ANP1(Desikan R et al., 1999), and AtMPK6 are involved in cold stress as we knew before. Additionally, in tobacco, under $H_2O_2$ and ozone treatment, the ortholog of AtMPK, SIPK1 will be activated as well (Samuel MA et al., 2000). These findings imply that multiple MAPK modules mediate oxidative stress responses and that MAPK cascades are not only induced by ROS but may also regulate ROS levels. Meanwhile, activating SIPK1 (salicylic acid-induced protein kinase), who is an NO-activated protein kinase in tobacco, can not process without SA, which brings a suggestion for the existence of cross-talk between ROS, hormone signaling and MAPKs. Here is some other evidence coming from studies on stomatal movement (Eckardt NA., 2009). In guard cells of Vicia faba, MAP2K is believed to regulate stomatal movement through mediating $H_2O_2$ generation induced by ABA (Song XG et al., 2008). Later, in guard cells studies, MAPK9 and MAPK12 have been proved to serve as positive regulators acting downstream of reactive oxygen species and calcium signaling in ABA signaling. In 2.2.4, we talk about ABA- and $Ca^{2+}$-induced stomatal closure, so we can't help wonder is there any link between MAPK and ABA- and $Ca^{2+}$-induced pathways? Yes, it has been found that ABA and $Ca^{2+}$ signals cannot activate anion channels in mpk9/12 mutants, thus indicating that these two MAPKs act between the ABA and $Ca^{2+}$ signals and the anion channels. (Jammes F et al., 2009).

Even if the role of MAPK in ABA signaling has not yet been directly confirmed, the attempts to clarify the relationship between hormone and MAPKs never stop. Not just for ABA, the role of MAPK signaling cascades in auxin signaling and ethylene biosynthesis has been documented in numerous studies (Dai Y et al., 2006; Xu J et al., 2008). Eventually, the genes in hormone biosynthesis (ethylene) and responses (auxin) are altered, it is of great significance to distinguish the direct targets of MAPK cascade from those that are regulated by altered hormonal and oxidative stress responses. On the other hand, recently in heavy metal stress studies, we find data supporting that MAPK3 and MAPK6 are activated responding to cadmium through ROS accumulation produced by oxidative stress in Arabidopsis (Liu XM et al., 2010), which further demonstrate the close relevance between ROS and MAPKs.

Meanwhile, by searching for the linkage between MAPKs and its various substrates involving other kinases and transcription factors, we gain much more information about MAPK cascades. Over-expression of MAP2K2 affects genes for several transcription factors (such as RAV1, STZ, ZAT10, ERF6, WRKY, and CBF2), disease resistance proteins, cell wall related proteins, enzymes involved in some secondary metabolisms and an 1-aminocyclopropane-1-carboxylic acid synthase (ACS). In the case of ACS, the rate-limiting enzyme of ethylene biosynthesis, the phosphorylation by MAPKs and by CDPKs affects protein stability and turnover, which again shows us the complicated cross-talk network (Bernhard Wurzinger et al., 2011). So it is a big challenge to identifying the targets of MAPK cascades, but the researches on other protein kinases leave us useful clue to find the answer.

### *2.4.2. Other protein kinases*

Whether at the transcript level or activity level, protein kinases are induced by a variety of abiotic stress, which indicates their powerful participation in signaling process. Moreover no matter suppressing or overexpressing these kinases, there both exists data showing that in transgenic plants, stress responses has changed. So far, we know several protein kinases involved in stress tolerance are stimulated by ABA, such as most of SNF1-related kinases (SnRKs) like SnRK2, SnRK3 (CIPK), CDPK and MAPK families. But others like Glycogen synthase kinase 3 (GSK3) ( Jonak and Hirt, 2002; Koh et al., 2007), S6 kinase (S6K) (Mahfouz et al., 2006), SERK (Marcelo O. Santos et al., 2009) also attract a lot of attention.

Now let's start from SnRKs. The SnRK2 family members are plant-specific kinases relating to abiotic stresses responses and abscisic acid (ABA)-dependent plant development. They have been classed into three groups; members of group 1 are not activated by ABA, and group 2 also will not be activated or weekly activated, while group 3 is strongly activated by ABA. In *Arabidopsis* the SnRK2 subfamily consists of 10 members. Except SnRK2.9, all SnRK2s are activated by osmotic and salt stress (Boudsocq, M. et al., 2004). Take SnRK2.6 (OST1) as an example. SnRK2.6 functions in the ABA signaling pathway upstream ABA-induced ROS production. It is related to the ABA-activated protein kinase AAPK in Vicia faba and also associates with SNF1 protein kinase. NADPH oxidases function in ABA signal transduction, also targeted by the SnRK2.6 kinase (Nakashima et al., 2010). Generally speaking, regulating the response to ABA through SnRK2s pathways is to directly phosphorylate various downstream targets such as ion channels (SLAC1, KAT1) and ABFs and other specific TFs required for expression of stress-responsive genes (Anna Kulik et al., 2011). By the way, the SnRK2 subfamily is conserved in land plants. No wonder their role in ABA signaling and osmotic stress responses have also been found in pea, barley, rice and *zea mays* (Shen, Q. et al., 2001; Kobayashi et al., 2004; Huai, J. et al., 2008). As for SnRK3 (CIPK), we compact it here that CIPK1/3/8/14/15/20/23/24 take part in ABA signaling. CIPKs play a main role in plant ion homeostasis and abiotic stress tolerance by regulating $H^+$, $Na^+$, $Ca^{2+}$ and $NO_3^-$ transporters and $K^+$ channels and interacting with TFs (Kudla, J. et al., 2010). Moreover, the CDPKs that are involved in ABA signaling are CPK3/4/6/11/32. CPK4 and CPK11 are closely related genes and both phosphorylate the transcription factors ABF1 and ABF4. Based on a majority of evidence, it can be concluded that CDPKs target core ABA signaling components (Geiger, et al., 2010). A role for MAPKs in ABA signaling has been shown above, but posted here again (MPK1/2/3/6/9/12). Taken together, none of these protein kinase families function specifically in ABA signaling, which still need us to stress on functional redundancy and complicated cross-talk network in the future research.

As follows, we take a brief look at other kinases. CDKs (Cyclin dependent protein kinases) are a large family of serine/threonine protein kinases and mainly function in ensuring that cells progress in order over the different cell division stages. But their roles in abiotic stress responses turn out to be more eye-catching, which can be reflected in heat, cold, drought and salt stimuli researches (reviewed by Georgios Kitsios, 2011). Other findings like Somatic Embryogenesis Receptor Kinase (SERK) relating to somatic embryogenesis and apomixis (Marcelo O. Santos et al., 2009), AtNEK6, a member of the NIMA (never in mitosis A)-related kinases (NEKs) (Lee SJ et al., 2010), bring us a large amount of information to explore the functional importance of various kinases in abiotic stress.

### 2.4.3. *Protein phosphatases*

During phosphorylation process, the job for protein phosphatases is to remove the phosphate added by protein kinases. Based on their substrate specificity, protein phosphatases can be classified, at least three families, as PPP family and PPM family composed by serine/threonine phosphatases, PTP family comprising tyrosine phosphatases, and dual specificity phosphatases (dsPTPs/dsPPase). As the largest group of phosphatases in plants, serine/threonine phosphatases can be further divided into PP1, PP2A, PP2B, and PP2C. For stress signal transduction, involvement of PP2C, PP2A, PTP, dsPPase have been reported in ABA or stress signal transduction, but the best-known example is the PP2C. On the other hand, many experiments have found that the relationship between MAPKs and phosphatases both exists in animals and plants. It has been shown that tyrosine-specific phosphorylation is associated with plant MAPKs, which again demonstrate the essential position phosphatases take in signaling pathways.

For the biggest branch of protein phosphatases in plants, in Arabidopsis alone, 76 PP2C genes have been identified early (Kerk et al. 2002). Further researches say ABA and other abiotic stress stimuli can induce A-type PP2C expression. Furthermore the A-type PP2C phosphatases, ABI1, ABI2, HAB1 (P2CHA) and PP2CA (AHG3) have been proved to directly interact with PYR/RCAR ABA receptors, however they always act as negative regulators at different layers of ABA signaling. (Merlot et al., 2001; Umezawa et al., 2009; Nishimura et al., 2010).

Working on other phosphatases like PP1, we find data in *Vicia faba* studies supporting its involvement in stomatal opening during the response to blue light (Takemiya et al. 2006). As to PP2A, five genes encoding its catalytic subunit has been identified in rice, and three of them show expression alteration under abiotic stress (Yu et al. 2003, 2005). And the activity of PP2B needs the help from calcium (Luan 2003). Some others like DSP4, a dual-specificity phosphatase, has been demonstrated to bind starch and interact with AKIN11, a SNF 1-related kinase in Arabidopsis (Fordham-Skelton et al. 2002; Kerk et al. 2006). Anyway we do need more facts to clarify the function network between phosphatases and other signals like hormone, ion channels, kinases actions.

Finally, we can see that the biggest challenge for abiotic stress signal research is to elaborate the network of kinases and phosphatases, and their relationship with a wide range of substrates, as well as understand the phosphorylation states and how phosphorylation-dependent activity culminates in the process of coping with a particular environmental stress.

## 2.5. The role of TFs and genes in certain abiotic stress situations

By the end of signaling, the biggest assignment is to temporally and spatially regulate stress-induced genes expression, and it is almost done at transcriptional level(Rushton and Somssich 1998). Obviously the most contributing workers are transcription factors (TFs) who modulate genes expression by binding to specific DNA sequences in the promoters of target genes (Chaves and Oliveira 2004). Thanks to coordination between TFs and genes, new transcripts are synthesized and stress adaptations have been realized within just a few hours. Consequently, that is why TFs are such a group of powerful targets being so popular in genetic engineering field aiming at improving stress resistance in crop plants. But, who are they? While, most of them belong to several big families, naming AP2/ERF (ethylene responsive element binding factor), Zn finger, basic leucine zipper (bZIP), basic helix-loop-helix (bHLH), WRKY, MYB, and NAC. Indeed we frequently mentioned them in formal description. So let's start to wander in their complicated regulatory network designed for the also complex stress situations.

### *2.5.1. Drought*

Based on recent data, the rate of land area experiencing drought is uprising and probably goes up to 30% by the end of this century (Yi et al. 2010). And that is a big threaten for plants lives, for drought is very likely to give rise to arrest of photosynthesis, disturbance of metabolism and finally plant death (Jaleel et al. 2008). But plants will react efficiently. With a few second of water loss, phosphorylation status of a protein will be triggered and later when the suffering time reached hours or days long, gene expression and plant morphology occur (Verslues and Bray, 2006). By researching on Arabidopsis plants under water-deficit stress, more than 800 induced genes have been identified (Bray, 2004), who play key roles in signal transduction, transcriptional regulation, cellular metabolism and transport, as well as cellular structures protection. Meanwhile, when water deficit comes two major transcriptional regulatory pathways of gene expression show up. One group is TFs called dehydration response element binding protein (DREB) coming from AP2/ERF family and another is NAC, AREB/ABF, WRKY and MYB group. The former work in the ABA-independent pathway, latter known to be as the most responsive to ABA signaling under drought.

The DREB proteins contain an ERF/AP2 DNA-binding domain that is quite conserved. The TFs containing it are widely found in many plants, including Arabidopsis (Okamuro et al. 1997 ), tomato (Zhou et al. 1997 ), tobacco (Ohme-Takagi and Shinshi 1995 ), rice (Weigel 1995 ) and maize (Moose and Sisco 1996 ). And a conserved Ser/Thr-rich region next to ERF/AP2 domain is considered to be responsible for DREB proteins phosphorylation (Liu et al., 1998). Latter, Agarwal et al. found high sequence similarity exists in different DREB proteins by amino acid alignment analysis (Agarwal et al. , 2006).

Generally, The DREB TFs could be divided into DREB1 and DREB2, and they participate in signal transduction pathways under low temperature and dehydration respectively. They belong to plants-distinctive ERF family. ERF proteins share a conserved domain binding to the C-repeat CRT/dehydration responsive element (DRE) motif engaged in the expression of cold and dehydration responsive genes (Agarwal et al., 2006). The expression of DREB2A and its homolog DREB2B are induced by dehydration and high salt stress (Liu et al., 1998; Nakashima et al., 2000 ). Furthermore, the expression of DREB genes is induced by abiotic stresses at different time periods (Liu et al. 1998; Dubouzet et al. 2003). But for the character of tissue-specific expression, the data is still in short.

For the ABA-dependent gene induction under water-deficit condition, we want to stress on another group of TFs. Firsly, MYB, MYC, homeodomain TFs and a family of transcriptional repressors (Cys2/His2-type zinc-finger proteins) are involved in the ABA response to water deficit. The promoter region of Responsive to Dehydration 22 (RD22), who is induced by ABA, contains MYC and MYB *cis*-element recognition sites. And MYC and MYB TFs only accumulate after an increase of ABA concentration, and Over-expressing these TFs lead to promoted ABA sensitivity and drought tolerance (Abe et al. 2003). Also other data indicate that two R2R3-MYB TFs (AtMYB60 and AtMYB61) are directly involved in stomatal dynamics in Arabidopsis regulated by light conditions, ABA and water stress (Liang et al. 2005). Very recently, in Arabidopsis, it has been strongly suggested that WRKY TFs possibly act downstream of at least two ABA receptors, the cytoplasmic PYR/PYL/RCAR protein phosphatase 2C-ABA complex and the chloroplast envelope–located ABAR–ABA complex. And the promoter-binding experiments show that the target genes for WRKY TFs involved in ABA signalling include *ABF2/4, ABI4/5, MYB2, DREB1a/2a, RAB18, RD29A* and *COR47*. Other findings in a large sense prove us ome WRKY TFs are positive regulators of ABA-mediated stomatal closure being relevant with drought responses (reviewed by Deena L. Rushton et al., 2011). On the other hand, it has been realized earlier that the ABA response element (ABRE) is bound by basic Leucine Zipper Domain (bZIP-type) TFs, and three Arabidopsis bZIP TFs (AREB1/ABF2, AREB2/ABF4, and ABF3) are activated through phosphorylation reacting to water deficit and ABA treatment. Other NAC domain proteins ANAC019, ANAC055, and ANAC072 are also induced by the same treatment. And in guard cells, it has been shown that the strong induction of Stress Responsive–NAC1 (SNAC1) gene expression by drought also affect stomatal closure (Hu et al. 2006).

### *2.5.2. Flooding stress*

Flooding and submergence are two stresses lead to anoxic conditions in the root system. Under this stress condition, both anoxia and hypoxia are defined by $O_2$ shortage. But diverse plants have their own way to adjust to it. Some lowland rice cultivars, such as FR13A, can survive submergence by suppressing shoot elongation. At the molecular level, Submergence-1 (Sub1), which is derived from FR13A, is a major quantitative trait locus contributing to great submergence tolerance (Xu et al. 2006). And three sequentially arrayed genes (designated *Sub1A, Sub1B*, and *Sub1C*) has been identified. *Sub1A* has been proved to encode an ERF domain–containing TF associated with the induction of low oxygen escape syndrome (LOES) (Bailey-Serres and Voesenek, 2008 ).

In plants, different cell types exhibit a conserved response to low oxygen levels at the molecular level (Mustroph et al. 2010). This response includes the induction of genes after 30 min under hypoxia, whose expression is maintained for several hours (Klok et al. 2002; van Dongen et al. 2009). The increased transcript levels of these genes are further accompanied by active combination between mRNAs and polysomes reflecting promoted translation process (Branco-Price et al. 2008). Unfortunately, in plants the mechanism by which oxygen is perceived has not been clarified. But researches on hypoxia-responsive TFs can help us a lot to investigate the regulation of the hypoxic response. Usually these TFs are detected in families like MYB, NACs [Arabidopsis Transcription Activation Factor (ATAF) and Cup-shaped Cotyledons (CUC)], Plant Homeodomain (PHD) and ERF families (Hoeren et al. 1998; Christianson et al. 2009; Licausi et al. 2010b). And on the other hand, Microarray data in Arabidopsis and rice research find us several transcription factors whose expression increases induced by oxygen deprivation, such as heat shock factors, MADS-box proteins, and WRKY factors (Lasanthi-Kudahettige et al., 2007). Recently Licausi et al. (2010a) have identified TFs that are differentially expressed under hypoxic conditions. The results indicate members of the AP2 / ERF-type family are the most common upregulated TFs, followed by Zinc-finger and basic helix-loop-helix (bHLH-type) TFs. TFs belonging to the bHLH family also appear in the downregulated part, together with members from the bZIP and MYB families.

On the other hand, by *silico* experiments and trans-activation assays it has been confirmed that five hypoxia-induced TFs (At4g29190; LBD41, At3g02550; HRE1, At1g72360; At1g69570; At5g66980) from different TF families [Zinc Finger, Ligand Binding Domain (LBD)/Lateral Organ Boundary Domain, ERF, DNA binding with one finger (DOF), ARF] respectively showed the ability to regulate the expression of hypoxia responsive genes (Licausi et al. 2010b). Other evidence relating to TFs and adaptive response to low oxygen will refer to redox-sensitive TFs. ZAT12, a putative zinc finger-containing TF, is identified as an important link in the oxidative stress response signalling network in Arabidopsis (Rizhsky et al. 2004), for its transcript levels were remarkably mounted up in response to hypoxia and anoxia in several independent analyses (Branco-Price et al. 2005).

### 2.5.3. *Salinity*

Plants growing in high salt concentrations, as we know, will suffer from osmotic stress and take actions like closing stomata and reducing cell expansion in young leaves and root tips. Subsequently, accumulation of ions, especially sodium ($Na^+$), in the photosynthetic tissues, will hit photosynthetic components such as enzymes, chlorophylls, and carotenoids (Davenport et al. 2005), followed by secondary stresses (oxidative stress and nutritional disorders) (Hasegawa et al. 2000; Chinnusamy et al. 2006).

One of the main strategies taken to improve plant salt tolerance is to re-establish ion homeostasis by counteracting the osmotic component of the stress to avoid toxic concentrations within the cytoplasm (Munns and Tester 2008). Recently, in Arabidopsis it has been confirmed that the effective establishing and maintaining ion homeostasis is mediated mainly by a Salt Overly Sensitive (SOS) signal pathway, which we refer to in $Ca^{2+}$ part. Recently, SOS4 and SOS5 have also been characterized (Mahajan et al. 2008). Similar to *SOS1*, *Arginine Vasopressin1* (*AVP1*) and A. thaliana *Na⁺/H⁺ exchanger1* (*AtNHX1*) genes contribute to ion homeostasis also (Gaxiola et al. 2001; Zhang et al. 2001). Besides these genes, overexpression of genes encoding LEA proteins, such as the barley HVA1 (Xu et al. 1996) and wheat dehydrin-5 (DHN-5) (Brini et al. 2007), is confirmed to be able to enhance plant salt tolerance. And regulating Lea gene expression is mediated by both ABA dependent and independent signalling pathways, both of which use $Ca^{2+}$ signaling to induce Lea gene expression during salinity.

Meanwhile, what do TFs do in salt stress signaling? Based on researches on Cor/Lea salinity stress responsive genes, whose expression is mediated $Ca^{2+}$and ABA in salt stress signaling, it has been indicated that various upstream TFs will activate DRE/CRT, ABREs, MYC recognition sequence (MYCRS) and MYB recognition sequence (MYBRS) *cis*-elements. Also, with or without the involvement of ABA makes them differs from each other. On one hand, ABRE and MYB/MYC element-controlled gene expression is ABA-dependent, which activates bZIP TFs called AREB binding to ABRE element to induce the stress responsive gene (*RD29A*). However, even if ABA makes these TFs function in their own regulating ways, it has been shown that ABA-dependent and -independent TFs may also cross talk to each other in a synergistic way to amplify the response and improve stress tolerance (reviewed by Dortje Golldack et al., 2011).

### 2.5.4. *Extremes of temperature*

Signals of extreme temperature, from freezing to scorching, is perceived by membrane and transduced by different transduction components results in transcription of several genes. Cold stress directly inhibit metabolic reactions and indirectly produce harm through cold-induced osmotic prevents the expression of full genetic potential of plants owing to its direct inhibition of metabolic reactions and, indirectly, through cold-induced osmotic, oxidative and other stresses such as water uptake barriers caused by chilling and cellular dehydration induced by freeze. Cold stress, based on temperature range, are defined as chilling (<20°C) and/or freezing (<0°C) temperatures, both of which hurt plants in different ways. The former leads to slow biochemical reactions related to enzymes and membrane transport activities, while the latter forming ice crystal can cause membrane system disruption (Chinnusamy et al. 2007).

Numerous TFs in cold stress circumstances have been identified in Arabidopsis, homologs of which have been reported in other plants also. Significant progress has been made in the past decade in elucidating the transcriptional networks regulating cold acclimation. Firstly, AP2/ERF family TFs, CBFs play essential role in controlling genes in phosphoinositide metabolism, osmolyte biosynthesis, ROS detoxification, hormone metabolism and signalling (Lee et al. 2005). They can bind to *cis*-elements in the promoters of COR genes to activate gene expression. Earlier it was proved that *DREB1A/CBF3*, *DREB1B/CBF1* and *DREB1C/CBF2* genes, lying tandemly in the Arabidopsis genome, are induced by cold but not by dehydration or high salinity (Shinwari et al. 1998). *CBF* genes showed high expression under low temperature treatment and the transcript was detectable after 30 min exposed to 4 °C and reached maximum expression at 1 h (Medina et al. 1999). However this time-linking phenomenon differs from various plants. In rice, detecting *OsDREB1A* and *OsDREB1B* transcript might need to wait 40 min after cold exposure. By the way, CBF pathway might also have a crucial role in constitutive freezing tolerance (Hannah et al. 2006).

Moreover for many studies, emphasize has been laid on the ICE-CBF-COR transcriptional cascade. In Arabidopsis, ICE1 (Inducer of CBF Expression1), a MYC-type bHLH TF, can bind to MYC recognition elements in the *CBF3* promoter affecting its expression during cold acclimation (Chinnusamy et al. 2003). Besides being the inducer of *CBFs* transcription, ICE1 is also a transcriptional inducer of ZAT12, NAC072 and HOS9 in Arabidopsis (Benedict et al. 2006). Furthermore, by studying on ice1 mutation under cold stress, the genes in calcium signaling, lipid signaling or encoding receptor-like protein kinases are found to be affected (Lee et al., 2005). In conclusion, there do exists network between these components in cold signaling. Constitutive expressed ICE1 is actived through sumoylation and phosphorylation induced by cold stress, which then induce the transcription of *CBFs* and reprime *MYB15*. For *CBFs* expression, CBF2 acts as a negative regulator of *CBF1* and *CBF3* expression. Meanwhile, the expression of *CBFs* is negatively regulated by upstream TF MYB15 and ZAT12(Chinnusamy et al. 2007).

## 3. Conclusion

Thanks to so much efforts made by researchers in various fields, we have a pretty clear idea about how to develop our researching methods in abiotic stress field. That is exactly what we should concentrate on in the future. For plants' reactions to different kinds of stress, we must make more efforts in taking measures in focusing on systematic studies which so far can be taken as the best way to figure out what plants will do under certain circumstances.

As we all know, it is of incredibly great importance to understand more about abiotic stress which impacts a lot on plants, which will not only change our understanding of current environment we live, but also bring a plenty of benefits for improving human beings living standards. That is why at this part we hope we can get everyone's attention to how to explore plants kingdom and develop researches in a systematic way.

## Author details

Yin Gong

*Key Laboratory of Tropical Forest Ecology, Xishuangbanna Tropical Botanical Garden, Chinese Academy of Sciences, Kunming, Yunnan, China*

*College of Bioscience and Biotechnology, Hunan Agricultural University, Changsha, Hunan, China*

Liqun Rao

*College of Bioscience and Biotechnology, Hunan Agricultural University, Changsha, Hunan, China*

Diqiu Yu *

*Key Laboratory of Tropical Forest Ecology, Xishuangbanna Tropical Botanical Garden, Chinese Academy of Sciences, Kunming, Yunnan, China*

## 4. References

[1] Cooke DT, Munkonge FM, Burden RS, James CS. Fluidity and lipid composition of oat and rye shoot plasma membrane: effect of sterol perturbation by xenobiotics. Biochim Biophys Acta, 1991, 1061(2):156-62.

[2] Elkahoui S, Smaoui A, Zarrouk M, Ghrir R, Limam F. Salt-induced lipid changes in Catharanthus roseus cultured cell suspensions. Phytochemistry, 2004, 65(13):1911-7.

[3] Bewell MA, Maathuis FJ, Allen GJ, Sanders D. Calcium-induced calcium release mediated by a voltage-activated cation channel in vacuolar vesicles from red beet. *FEBS letters,* 1999, 458(1):41-4.

[4] Knight BW, Omurtag A, Sirovich L. The approach of a neuron population firing rate to a new equilibrium: An Exact Theoretical Result. *Neural Computation,* 2000, 12(5):1045-55.

[5] Sangwan V, Foulds I, Singh J, Dhindsa RS. Cold-activation of *Brassica napus BN115* promoter is mediated by structural changes in membranes and cytoskeleton, and requires $Ca^{2+}$ influx. *The Plant Journal,* 2001, 27(1):1-12

[6] Wang QY, Nick P. Cold acclimation can induce microtubular cold stability in a manner distinct from abscisic acid. *Plant Cell Physiology,* 2001, 42(9):999-1005.

[7] Plieth C. Temperature sensing by plants: calcium-permeable channels as primary sensors—A Model. *The Journal of Membrane Biology,* 1999, 172(2):121-7.

[8] Hasegawa PM, Bressan RA, Zhu J-K, Bohnert HJ. Plant cellular and molecular responses to high salinity. *Annual Review of Plant Physiology and Plant Molecular Biology,* 2000, 51(1):463-99.

[9] Suzuki I, Los DA, Kanesaki Y, Mikami K, Murata N. The pathway for perception and transduction of low-temperature signals in *Synechocystis*. *The EMBO Journal,* 2000, 19(6):1327-34.

* Corresponding Author

[10] Aguilar PS, Hernandez-Arriaga AM. Molecular basis of thermosensing: a two-component signal transduction thermometer in *Bacillus subtilis. The EMBO Journal,* 2001, 20(7):1681-91.

[11] Urao T, Yamaguchi-Shinozaki K, Shinozaki K. Two-component systems in plant signal transduction. *Trends in Plant Science,* 2000, 5(2):67-74.

[12] Maeda T, Wurgler-Murphy SM, Saito H. A two-component system that regulates an osmosensing MAP kinase cascade in yeast. *Nature.,* 1994, 369(6477):242-5.

[13] Urao T, Yakubov B, Satoh R, Yamaguchi-Shinozaki K. A transmembrane hybrid-type histidine kinase in Arabidopsis functions as an osmosensor. *The Plant Cell,* 1999, 11(9):1743-54.

[14] Zhu J-K. Plant salt tolerance. *Trends in Plant Science,* 2001, 6(2):66-71

[15] Xiong L, Zhu J-K. Molecular and genetic aspects of plant responses to osmotic stress. *Plant, Cell & Environment,* 2002, 25(2):131-9.

[16] Ullah H, Chen J-G, Young JC, Im K-H. Modulation of cell proliferation by heterotrimeric G protein in Arabidopsis. *Science,* 2001, 292 (5524): 2066-69.

[17] Wang X-Q, Ullah H, Jones AM, Assmann SM. G protein regulation of ion channels and abscisic acid signaling in Arabidopsis guard cells. *Science,* 2001, 292 (5524): 2070-72.

[18] Ma Y, Szostkiewicz I, Korte A. Regulators of PP2C phosphatase activity function as abscisic acid sensors. *Science,* 2009, 324 (5930): 1064-68.

[19] Park S-Y, Fung P, Nishimura N. Abscisic acid inhibits type 2C protein phosphatases via the PYR/PYL family of START proteins. *Science,* 2009, 324 (5930): 1068-71.

[20] Foyer CH and Noctor G. Ascorbate and glutathione: the heart of the redox hub. *Plant Physiology,* 2011, 155(1):2-18.

[21] Huchzermeyer B, Koyro HW. Salt and drought stress effects on photosynthesis. Handbook of Photosynthesis (2nd edition)., 2005, 12(2):145-51.

[22] Kant S, Kant P, Raveh E, Barak S. Evidence that differential gene expression between the halophyte, *Thellungiella halophila,* and *Arabidopsis thaliana* is responsible for higher levels of the compatible osmolyte proline and tight control of $Na^+$ uptake in *T. Halophila. Plant, Cell & Environment,* 2006, 29(7):1220-34.

[23] Türkan I, Demiral T. Recent developments in understanding salinity tolerance. Environmental and Experimental Botany, 2009, 67(1):2-9.

[24] Geissler B, Tungekar R, Satchell KJF. Identification of a conserved membrane localization domain within numerous large bacterial protein toxins. Proc Natl Acad Sci U S A., 2010, 107(12):5581-6.

[25] Vranova E, Atichartpongkul S. Comprehensive analysis of gene expression in *Nicotiana tabacum* leaves acclimated to oxidative stress. Proc Natl Acad Sci U S A., 2002, 96(16):10870-5.

[26] Foyer CH, Noctor G. Redox sensing and signalling associated with reactive oxygen in chloroplasts, peroxisomes and mitochondria. *Physiologia Plantarum,* 2003, 119(3):355-364.

[27] Moller IM. Plant mitochondria and oxidative stress: electron transport, NADPH turnover, and metabolism of reactive oxygen species. *Annual Review of Plant Physiology and Plant Molecular Biology,* 2001, 52(1):561-91.

[28] Rasmusson AG, Soole KL, Elthon TE. Alternative NAD (P) H dehydrogenases of plant mitochondria. *Annual Review of Plant Biology,* 2004, 55(1):23-39.

[29] Noctor G. Metabolic signalling in defence and stress: the central roles of soluble redox couples. *Plant, Cell & Environment,* 2006, 29(3):409-25.

[30] Rhoads DM, Umbach AL, Subbaiah CC, Siedow JN. Mitochondrial reactive oxygen species. Contribution to oxidative stress and interorganellar signaling. *Plant Physiology,* 2006, 141(2) 357-66.

[31] Pourrut B, Perchet G, Silvestre J, Cecchi M, Guiresse M, Pinelli E. Potential role of NADPH-oxidase in early steps of lead-induced oxidative burst in Vicia faba roots. Journal of Plant Physiology, 2008, 165(6):571-9.

[32] Leshem Y, Seri L, Levine A. Induction of phosphatidylinositol 3-kinase-mediated endocytosis by salt stress leads to intracellular production of reactive oxygen species and salt tolerance. *The Plant Journal,* 2007, 51(2):185-97.

[33] Foreman J, Demidchik V, Bothwell JH, Mylona P. Reactive oxygen species produced by NADPH oxidase regulate plant cell growth. Nature, 2003, 422(6930):442-6.

[34] Mori IC, Schroeder JI. Reactive Oxygen Species Activation of Plant $Ca^{2+}$ Channels. A Signaling Mechanism in Polar Growth, Hormone Transduction, Stress Signaling, and Hypothetically Mechanotransduction. *Plant Physiology,* 2004, 135(2):702-8.

[35] Gamaley IA, Klyubin IV. Roles of reactive oxygen species: signaling and regulation of cellular functions. International Review of Cytology, 1999, 188(1):203-55.

[36] Foyer CH, Noctor G. Redox homeostasis and antioxidant signaling: a metabolic interface between stress perception and physiological responses. *The Plant Cell,* 2005, 17(7):1866-75.

[37] Zhao L, Zhang F, Guo J, Yang Y, Li B, Zhang L. Nitric oxide functions as a signal in salt resistance in the calluses from two ecotypes of reed. *Plant Physiology,* 2004, 134(2):849-57.

[38] Prasad TK, Anderson MD, Martin BA. Evidence for chilling-induced oxidative stress in maize seedlings and a regulatory role for hydrogen peroxide. *The Plant Cell,* 1994, 6(1):65-74.

[39] Fadzilla NM, Finch RP, Burdon RH. Salinity, oxidative stress and antioxidant responses in shoot cultures of rice. Journal of Experimental Botany, 1997, 48(2):325-31.

[40] Lui J, Shono M. Characterization of mitochondria-located small heat shock protein from tomato (Lycopersicon esculentum). *Plant Cell Physiology,* 1999, 40(12): 1297-304.

[41] Lee BH, Won SH, Lee HS, Miyao M, Chung WI, Kim IJ. Expression of the chloroplast-localized small heat shock protein by oxidative stress in rice. Gene, 2000, 245(2):283-90.

[42] Moschou PN, Paschalidis KA, Delis ID. Spermidine exodus and oxidation in the apoplast induced by abiotic stress is responsible for H2O2 signatures that direct tolerance responses in tobacco. *The Plant Cell,* 2008, 20(6):1708-24.

[43] Zhang X, Zhang L, Dong F, Gao J, Galbraith DW. Hydrogen peroxide is involved in abscisic acid-induced stomatal closure in Vicia faba. *Plant Physiology,* 2001, 126(4):1438-48.

[44] Kwak JM, Mori IC, Pei ZM, Leonhardt N, Torres MA. NADPH oxidase AtrbohD and AtrbohF genes function in ROS-dependent ABA signaling in Arabidopsis. *The EMBO Journal,* 2003, 22(1):2623 -33.

[45] Guo FQ, Okamoto M, Crawford NM. Identification of a plant nitric oxide synthase gene involved in hormonal signaling. *Science,* 2003, 302(5642):100-3.

[46] Ortiz-Masia D, Perez-Amador MA, Carbonell J. Diverse stress signals activate the C1 subgroup MAP kinases of *Arabidopsis. FEBS letters,* 2007, 581(9):1834-40.

[47] Dóczi R, Brader G, Pettkó-Szandtner A. The Arabidopsis mitogen-activated protein kinase kinase MKK3 is upstream of group C mitogen-activated protein kinases and participates in pathogen signaling. *The Plant Cell,* 2007, 19(10):3266-79.

[48] Munnik T, Testerink C. Plant phospholipid signaling: "in a nutshell". *Journal of Lipid Research,* 2009, 50(2):260-5.

[49] Xue HW, Chen X, Mei Y. Function and regulation of phospholipid signalling in plants. *Biochemical Journal,* 2009, 421(2): 145–156.

[50] Munnik T, Vermeer JEM. Osmotic stress-induced phosphoinositide and inositol phosphate signalling in plants. *Plant, Cell & Environment,* 2010, 33(4):655-9.

[51] Testerink C, Munnik T. Phosphatidic acid: a multifunctional stress signaling lipid in plants.*Trends in Plant Science,* 2005, 10(8):368-75.

[52] Arisz SA, Testerink C, Munnik T. Plant PA signaling via diacylglycerol kinase. Biochimica et Biophysica Acta (BBA) - Molecular and Cell Biology of Lipids, 2009, 1791(9):869-75.

[53] Li M, Hong Y, Wang X. Phospholipase D-and phosphatidic acid-mediated signaling in plants. Biochimica et Biophysica Acta (BBA) - Molecular and Cell Biology of Lipids, 2009, 1791(9):927-35.

[54] Mishkind M, Vermeer JEM, Darwish E. Heat stress activates phospholipase D and triggers PIP2 accumulation at the plasma membrane and nucleus. *The Plant Journal,* 2009, 60(1):10-21.

[55] Testerink C, Munnik T. Molecular, cellular, and physiological responses to phosphatidic acid formation in plants. Journal of Experimental Botany, 2011, 62(7):2349-61.

[56] Zonia L, Munnik T. Osmotically induced cell swelling versus cell shrinking elicits specific changes in phospholipid signals in tobacco pollen tubes. *Plant Physiology,* 2004, 134(2):813-23.

[57] Darwish E, Testerink C, Khalil M. Phospholipid signaling responses in salt-stressed rice leaves. *Plant Cell Physiology,* 2009, 50 (5): 986-97.

[58] Hong Y, Zhang W, Wang X. Phospholipase D and phosphatidic acid signalling in plant response to drought and salinity. *Plant, Cell & Environment,* 2010, 33(4):627-35.

[59] Bargmann BOR, Laxalt AM, Riet B. Multiple PLDs required for high salinity and water deficit tolerance in plants. *Plant Cell Physiology,* 2009, 50 (1): 78-89.

[60] Mishra G, Zhang W, Deng F, Zhao J. A bifurcating pathway directs abscisic acid effects on stomatal closure and opening in Arabidopsis. *Science,* 2006, 312 (5771): 264-6.

[61] Arisz S, Munnik T. Diacylglycerol kinase. Lipid Signaling in Plants, 2010, 16(2010):107-14.

[62] Hirayama T, Ohto C, Mizoguchi T. A gene encoding a phosphatidylinositol-specific phospholipase C is induced by dehydration and salt stress in Arabidopsis thaliana. Proc Natl Acad Sci U S A., 1995, 92(9):3903-7.

[63] Sanchez JP, Chua NH. Arabidopsis PLC1 is required for secondary responses to abscisic acid signals. *The Plant Cell,* 2001, 13(5):1143-54.

[64] Popescu SC, Popescu GV, Bachan S, Zhang Z, Seay M, Gerstein M, Snyder M, Dinesh-Kumar SP. Differential binding of calmodulin-related proteins to their targets revealed through high-density Arabidopsis protein microarrays. Proc Natl Acad Sci U S A., 2007, 104(11):4730-5.

[65] Peters C, Li M, Narasimhan R, Roth M. Nonspecific phospholipase C NPC4 promotes responses to abscisic acid and tolerance to hyperosmotic stress in Arabidopsis. *The Plant Cell,* 2010, 22(8):2642-59.

[66] Testerink C, Dekker HL, Lim ZY, Johns MK. Isolation and identification of phosphatidic acid targets from plants. *The Plant Journal,* 2004, 39(4):527-36.

[67] Yu L, Nie J, Cao C, Jin Y, Yan M, Wang F, Liu J. Phosphatidic acid mediates salt stress response by regulation of MPK6 in Arabidopsis thaliana. *New Phytologist,* 2010, 188(3):762-73.

[68] Guo L, Wang X. Crosstalk between Phospholipase D and Sphingosine Kinase in Plant Stress Signaling. *Frontiers in plant science,* 2012, 51(3):3389-99.

[69] Lemtiri-Chlieh F, MacRobbie EAC. Inositol hexakisphosphate mobilizes an endomembrane store of calcium in guard cells. Proc Natl Acad Sci U S A., 2003, 100 (17): 10091-5.

[70] DeWald DB, Torabinejad J, Jones CA. Rapid accumulation of phosphatidylinositol 4, 5-bisphosphate and inositol 1, 4, 5-trisphosphate correlates with calcium mobilization in salt-stressed Arabidopsis. *Plant Physiology,* 2001, 126 (2): 759-69.

[71] Peters C, Li M, Narasimhan R, Roth M. Nonspecific phospholipase C NPC4 promotes responses to abscisic acid and tolerance to hyperosmotic stress in Arabidopsis. *The Plant Cell,* 2010, 22 (8): 2642-59.

[72] Scherer GFE, Labusch C, Effendi Y. Frontiers: Phospholipases and the Network of Auxin Signal Transduction with ABP1 and TIR1 as Two Receptors: A Comprehensive and Provocative Model. *Frontiers in Plant Physiology,* 2012, 56 (3): 652-63.

[73] Busch W, Benfey PN. Information processing without brains–the power of intercellular regulators in plants. *Development,* 2010, 137 (2): 1215-1226.

[74] Sachs T, Thimann KV. The role of auxins and cytokinins in the release of buds from dominance. American Journal of Botany, 1967, 54(1):136-44.

[75] Thimann KV, Skoog F. Studies on the growth hormone of plants: III. The inhibiting action of the growth substance on bud development. Proc Natl Acad Sci U S A., 1933, 19(7): 714–6.

[76] Sauter A, Davies WJ, Hartung W. The long-distance abscisic acid signal in the droughted plant: the fate of the hormone on its way from root to shoot. *Journal of Experimental Botany,* 2001, 52(363): 1991–7.

[77] Goda H, Sasaki E, Akiyama K. The AtGenExpress hormone and chemical treatment data set: experimental design, data evaluation, model data analysis and data access. *The Plant Journal,* 2008, 55(3): 526–42.

[78] Chapman EJ, Estelle M. Mechanism of auxin-regulated gene expression in plants. *Annual Review of Genetics,* 2009, 43(1): 265–85.

[79] Santner A, Estelle M. The ubiquitin-proteasome system regulates plant hormone signaling. *The Plant Journal,* 2010, 61(6): 1029–40.

[80] Bartoli CG, CasalonguéCA, Simontacchi M. Interactions between hormone and redox signalling pathways in the control of growth and cross tolerance to stress. Environmental and Experimental Botany, 2012, 52(3): 139–47.

[81] Cui J, Zhou Y, Ding JG. Role of nitric oxide in hydrogen peroxide-dependent induction of abiotic stress tolerance by brassinosteroids in cucumber. Plant, Cell & Environment, 2011, 34(2): 347–58.

[82] Schroeder JI, Kwak JM, Allen GJ. Guard cell abscisic acid signalling and engineering drought hardiness in plants. *Nature,* 2001, 410(6826): 327–30.

[83] Cutler SR, Rodriguez PL, Finkelstein RR. Abscisic acid: emergence of a core signaling network. Annual Review of Plant Biology, 2010, 61:651-79

[84] Ghanem ME, Hichri I, Smigocki AC, Albacete A. Root-targeted biotechnology to mediate hormonal signalling and improve crop stress tolerance. Plant Cell Reports, 2011, 30(5): 807–23.

[85] Argueso CT, Ferreir FJ. Environmental perception avenues: the interaction of cytokinin and environmental response pathways. *Plant, Cell & Environment,* 2009, 32(9): 1147–60.

[86] Eyidogan F, Oz MT, Yucel M, Oktem HA. Signal Transduction of Phytohormones Under Abiotic Stresses. Phytohormones and Abiotic Stress Tolerance in Plants, 2012, 3(642): 978–1007.

[87] Zhang SW, Li CH, Cao J, Zhang YC, Zhang SQ. Altered architecture and enhanced drought tolerance in rice via the down-regulation of indole-3-acetic acid by TLD1/OsGH3.13 activation. *Plant Physiology,* 2009, 151(4): 1889–901.

[88] Divi UK, Krishna P. Brassinosteroid: a biotechnological target for enhancing crop yield and stress tolerance. New Biotechnology, 2009, 26(3): 131–6.

[89] Sharp RE. Interaction with ethylene: changing views on the role of abscisic acid in root and shoot growth responses to water stress. *Plant, cell & environment,* 2002, 25(2): 211–22.

[90] Stepanova AN, Alonso JM. Ethylene signaling and response: where different regulatory modules meet. *Current Opinion in Plant Biology,* 2009, 12(5): 548–55.

[91] Jones B, Gunneras SA, Petersson SV. Cytokinin regulation of auxin synthesis in Arabidopsis involves a homeostatic feedback loop regulated via auxin and cytokinin signal transduction.*The Plant Cell,* 2010, 22(9): 2956–69.

[92] Wang L, Wang Z, Xu Y, Joo SH, Kim SK. OsGSR1 is involved in crosstalk between gibberellins and brassinosteroids in rice. *The Plant Journal,* 2009, 57(3): 498–510.

[93] Alonso-Ramirez A, Rodriguez D, Reyes D. Evidence for a role of gibberellins in salicylic acid-modulated early plant responses to abiotic stress in Arabidopsis seeds. *Plant Physiology,* 2009, 150(3): 1335–1344.

[94] Acharya BR, Assmann SM. Hormone interactions in stomatal function. Plant Molecular Biology, 2009, 69(4): 451–62.

[95] Divi UK, Rahman T, Krishna P. Brassinosteroid-mediated stress tolerance in Arabidopsis shows interactions with abscisic acid, ethylene and salicylic acid pathways. *BMC plant biology*, 2010,10(151):1186-471.

[96] Williamson RE, Ashley CC. Free Ca2++ and cytoplasmic streaming in the alga Chara. *Nature*, 1982, 296:647- 51

[97] Johnson CH, Knight MR, Kondo T, Masson. P Circadian oscillations of cytosolic and chloroplastic free calcium in plants. Science, 1995, 269(5232):1863-5.

[98] Tracy FE, Gilliham M, Dodd AN. NaCl-induced changes in cytosolic free Ca2+ in Arabidopsis thaliana are heterogeneous and modified by external ionic composition. *Plant, Cell & Environment*, 2008, 31(8):1063-73.

[99] Kudla J, Batistic O, Hashimoto K. Calcium signals: the lead currency of plant information processing. The Plant Cell, 2010, 22(3):541-63.

[100] Vinogradova MV, Reddy VS, Reddy ASN. Crystal structure of kinesin regulated by Ca2+-calmodulin. *The Journal of Biological Chemistry*, 2004, 279(1):23504-9.

[101] Allen GJ, Sanders D. Calcineurin, a type 2B protein phosphatase, modulates the Ca2+-permeable slow vacuolar ion channel of stomatal guard cells. *The Plant Cell*, 1995, 7(9):1473-83.

[102] Dodd AN, Kudla J, Sanders D. The language of calcium signaling. *Annual Review of Plant Biology*, 2010, 61(5):593-620.

[103] Snedden WA, Fromm H. Calmodulin as a versatile calcium signal transducer in plants. *New Phytologist*, 2001, 151(1):35-66.

[104] Hoeflich KP, Ikura M. Calmodulin in action: diversity in target recognition and activation mechanisms. Cell, 2002, 108(6):739-42.

[105] Yang T, Poovaiah BW. Calcium/calmodulin-mediated signal network in plants. Trends in Plant Science, 2003, 8(10):505-12.

[106] McCormack E, Tsai YC, Braam J. Handling calcium signaling: Arabidopsis CaMs and CMLs. Trends in Plant Scienc, 2005, 10(8):383-9.

[107] Popescu SC, Popescu GV, Bachan S, Zhang Z, Seay M, Gerstein M, Snyder M, Dinesh-Kumar SP. Differential binding of calmodulin-related proteins to their targets revealed through high-density Arabidopsis protein microarrays. Proc Natl Acad Sci U S A., 2007, 104(11):4730-5

[108] Cheng SH, Willmann MR, Chen HC, Sheen J. Calcium signaling through protein kinases. The Arabidopsis calcium-dependent protein kinase gene family. Plant Physiology, 2002, 129(2):469-85.

[109] Luan S, Kudla J, Rodriguez-Concepcion M, Yalovsky S, Gruissem W. Calmodulins and calcineurin B-like proteins: calcium sensors for specific signal response coupling in plants. Plant Cell, 2002, 14 Suppl:S389-400.

[110] Luan S, Lan W, Chul Lee S. Potassium nutrition, sodium toxicity, and calcium signaling: connections through the CBL-CIPK network. Current Opinion in Plant Biology, 2009, 12(3):339-46.

[111] Weinl S, Kudla J. The CBL-CIPK $Ca^{2+}$-decoding signaling network: function and perspectives. New Phytology, 2009, 184(3):517-28.
[112] Batistic O, Waadt R, Steinhorst L, Held K, Kudla J. CBL-mediated targeting of CIPKs facilitates the decoding of calcium signals emanating from distinct cellular stores. Plant Journal, 2010, 61(2):211-22.
[113] Clark GB, Roux SJ. Annexins of plant cells. Plant Physiology, 1995, 109(4):1133-9.
[114] Laohavisit A, Davies JM. Annexins. New Phytology, 2011,189(1):40-53.
[115] White PJ, Bowen HC, Demidchik V, Nichols C, Davies JM. Genes for calcium-permeable channels in the plasma membrane of plant root cells. Biochim Biophys Acta, 2002, 1564(2):299-309.
[116] McCormack E, Tsai YC, Braam J. Handling calcium signaling: Arabidopsis CaMs and CMLs. Trends in Plant Scienc, 2005, 10(8):383-9.
[117] DeFalco TA, Chiasson D, Munro K, Kaiser BN, Snedden WA. Characterization of GmCaMK1, a member of a soybean calmodulin-binding receptor-like kinase family. *FEBS letters*, 2010, 584(23):4717-24.
[118] Day IS, Reddy VS, Ali GS, Reddy ASN. Analysis of EF-hand-containing proteins in *Arabidopsis*. Genome Biology, 2002, 3(10): research0056.1–0056.24
[119] Harper JF, Harmon A. Plants, symbiosis and parasites: a calcium signalling connection. Nat Rev Mol Cell Biol., 2005, 6(7):555-66.
[120] Reddy AS, Reddy VS, Golovkin M. A calmodulin binding protein from Arabidopsis is induced by ethylene and contains a DNA-binding motif. Biochem Biophys Res Commun., 2000, 279(3):762-9.
[121] Park CY, Lee JH, Yoo JH, Moon BC, Choi MS, Kang YH, Lee SM, Kim HS, Kang KY, Chung WS, Lim CO, Cho MJ. WRKY group IId transcription factors interact with calmodulin. *FEBS letters*, 2005, 579(6):1545-50.
[122] Szymanski DB, Liao B, Zielinski RE. Calmodulin isoforms differentially enhance the binding of cauliflower nuclear proteins and recombinant TGA3 to a region derived from the Arabidopsis Cam-3 promoter. Plant Cell, 1996, 8(6):1069-77.
[123] Galon Y, Finkler A, Fromm H. Calcium-regulated transcription in plants. Molecular Plant, 2010, 3(4):653-69.
[124] BW Poovaiah, ASN Reddy. Calcium and signal transduction in plants. *Critical Reviews in Plant Sciences*, 1993, 12(3):185-211.
[125] Zhang Y, Xu S, Ding P, Wang D, Cheng YT, He J, Gao M, Xu F, Li Y, Zhu Z, Li X, Zhang Y. Control of salicylic acid synthesis and systemic acquired resistance by two members of a plant-specific family of transcription factors. Proc Natl Acad Sci USA., 2010, 107(42):18220-5.
[126] Kim SG, Kim SY, Park CM. A membrane-associated NAC transcription factor regulates salt-responsive flowering via FLOWERING LOCUS T in Arabidopsis. Planta, 2007, 226(3):647-54.
[127] Yoon HK, Kim SG, Kim SY, Park CM. Regulation of leaf senescence by NTL9-mediated osmotic stress signaling in Arabidopsis. Molecules and Cells, 2008, 25(3):438-45.

[128] Kim YS, Kim SG, Park JE, Park HY, Lim MH, Chua NH, Park CM. A membrane-bound NAC transcription factor regulates cell division in Arabidopsis. *The* Plant Cell, 2006, 18(11):3132-44.

[129] Kushwaha R, Singh A, Chattopadhyay S. Calmodulin7 plays an important role as transcriptional regulator in Arabidopsis seedling development. *The* Plant Cell, 2008, 20(7):1747-59.

[130] Dai X, Xu Y, Ma Q, Xu W, Wang T, Xue Y, Chong K. Overexpression of an R1R2R3 MYB gene, OsMYB3R-2, increases tolerance to freezing, drought, and salt stress in transgenic Arabidopsis. Plant Physiol*ogy*, 2007, 143(4):1739-51.

[131] Geisler M, Axelsen KB, Harper JF, Palmgren MG. Molecular aspects of higher plant P-type $Ca^{2+}$-ATPases. Biochim Biophys Acta, 2000, 1465(1-2):52-78.

[132] Munns R, James RA, Läuchli A. Approaches to increasing the salt tolerance of wheat and other cereals. Journal of Experimental Botany, 2006, 7(5):1025-43.

[133] Goldgur Y, Rom S, Ghirlando R, Shkolnik D, Shadrin N, Konrad Z, Bar-Zvi D. Desiccation and zinc binding induce transition of tomato abscisic acid stress ripening 1, a water stress- and salt stress-regulated plant-specific protein, from unfolded to folded state. Plant Physiol*ogy*, 2007, 143(2):617-28.

[134] Rabbani MA, Maruyama K, Abe H, Khan MA, Katsura K, Ito Y, Yoshiwara K, Seki M, Shinozaki K, Yamaguchi-Shinozaki K. Monitoring expression profiles of rice genes under cold, drought, and high-salinity stresses and abscisic acid application using cDNA microarray and RNA gel-blot analyses. Plant Physiol*ogy*, 2003, 133(4):1755-67.

[135] Liu J, Ishitani M, Halfter U, Kim CS, Zhu JK. The Arabidopsis thaliana SOS2 gene encodes a protein kinase that is required for salt tolerance. Proc Natl Acad Sci USA., 2000, 97(7):3730-4.

[136] Gong D, Guo Y, Schumaker KS, Zhu JK. The SOS3 family of calcium sensors and SOS2 family of protein kinases in Arabidopsis. Plant Physiol*ogy*, 2004, 134(3):919-26.

[137] Zhu JK. Regulation of ion homeostasis under salt stress. Current Opinion in Plant Biol*ogy*, 2003, 6(5):441-5.

[138] Brini F, Hanin M, Mezghani I, Berkowitz GA, Masmoudi K. Overexpression of wheat $Na^+/H^+$ antiporter TNHX1 and $H^+$-pyrophosphatase TVP1 improve salt- and drought-stress tolerance in Arabidopsis thaliana plants. Journal of Experimental Botany, 2007, 8(2):301-8.

[139] Perruc E, Charpenteau M, Ramirez BC, Jauneau A, Galaud JP, Ranjeva R, Ranty B. A novel calmodulin-binding protein functions as a negative regulator of osmotic stress tolerance in Arabidopsis thaliana seedlings. Plant *Journal*, 2004, 38(3):410-20.

[140] Yamaguchi-Shinozaki K, Shinozaki K. Transcriptional regulatory networks in cellular responses and tolerance to dehydration and cold stresses. *Annual Review of Plant Biology*, 2006, 57:781-803.

[141] Lecourieux D, Ranjeva R, Pugin A. Calcium in plant defence-signalling pathways. New Phytologist, 2006, 171(2):249-69.

[142] Saijo Y, Hata S, Kyozuka J, Shimamoto K, Izui K. Over-expression of a single $Ca^{2+}$-dependent protein kinase confers both cold and salt/drought tolerance on rice plants. Plant Journal, 2000, 23(3):319-27.

[143] Mishra NS, Tuteja R, Tuteja N. Signaling through MAP kinase networks in plants. Arch Biochem Biophys., 2006, 452(1):55-68.

[144] Dan I, Watanabe NM, Kusumi A. The Ste20 group kinases as regulators of MAP kinase cascades. Trends Cell Biol., 2001, 11(5):220-30.

[145] Whitmarsh AJ, Cavanagh J, Tournier C, Yasuda J, Davis RJ. A mammalian scaffold complex that selectively mediates MAP kinase activation. Science, 1998, 281(5383):1671-4.

[146] Rodriguez MC, Petersen M, Mundy J. Mitogen-activated protein kinase signaling in plants. *Annual Review of Plant Biology,* 2010, 61:621-49.

[147] Jonak C, Okrész L, Bögre L, Hirt H. Complexity, cross talk and integration of plant MAP kinase signalling. Current Opinion in Plant Biology, 2002, 5(5):415-24.

[148] Xiong L, Yang Y. Disease resistance and abiotic stress tolerance in rice are inversely modulated by an abscisic acid-inducible mitogen-activated protein kinase. Plant Cell, 2003, 15(3):745-59.

[149] Raman M, Chen W, Cobb MH. Differential regulation and properties of MAPKs. Oncogene, 2007, 26(22):3100-12.

[150] Taj G, Agarwal P, Grant M, Kumar A. MAPK machinery in plants: recognition and response to different stresses through multiple signal transduction pathways. Plant Signal Behav., 2010, 5(11):1370-8.

[151] Sinha AK, Jaggi M, Raghuram B, Tuteja N. Mitogen-activated protein kinase signaling in plants under abiotic stress. Plant Signal Behav., 2011, 6(2):196-203.

[152] Kovtun Y, Chiu WL, Tena G, Sheen J. Functional analysis of oxidative stress-activated mitogen-activated protein kinase cascade in plants. Proc Natl Acad Sci USA., 2000, 97(6):2940-5.

[153] Shou H, Bordallo P, Wang K. Expression of the Nicotiana protein kinase (NPK1) enhanced drought tolerance in transgenic maize. J Exp Bot., 2004, 55(399):1013-9.

[154] Ichimura K, Mizoguchi T, Yoshida R, Yuasa T, Shinozaki K. Various abiotic stresses rapidly activate Arabidopsis MAP kinases ATMPK4 and ATMPK6. Plant Journal, 2000, 24(5):655-65.

[155] Teige M, Scheikl E, Eulgem T, Dóczi R, Ichimura K, Shinozaki K, Dangl JL, Hirt H. The MKK2 pathway mediates cold and salt stress signaling in Arabidopsis. Molecular Cell., 2004, 15(1):141-52.

[156] Desikan R, Clarke A, Atherfold P, Hancock JT, Neill SJ. Harpin induces mitogen-activated protein kinase activity during defence responses in Arabidopsis thaliana suspension cultures. Planta, 1999, 210(1):97-103.

[157] Samuel MA, Miles GP, Ellis BE. Ozone treatment rapidly activates MAP kinase signalling in plants. Plant Journal, 2000, 22(4):367-76.

[158] Eckardt NA. Negative regulation of stress-activated MAPK signaling in Arabidopsis. Plant Cell, 2009, 21(9):2545.

[159] XG Song, XP She, LY Guo, ZN Meng, AX Huang. MAPK Kinase and CDP Kinase Modulate Hydrogen Peroxide Levels during dark-induced Stomatal Closure in Guard Cells of Vicia faba. Botanical Studies, 2008, 49(4):323-34.

[160] Jammes F, Song C, Shin D, Munemasa S, Takeda K, Gu D, Cho D, Lee S, Giordo R, Sritubtim S, Leonhardt N, Ellis BE, Murata Y, Kwak JM. MAP kinases MPK9 and MPK12 are preferentially expressed in guard cells and positively regulate ROS-mediated ABA signaling. Proc Natl Acad Sci USA., 2009, 106(48):20520-5.

[161] Dai Y, Wang H, Li B, Huang J, Liu X, Zhou Y, Mou Z, Li J. Increased expression of MAP KINASE KINASE7 causes deficiency in polar auxin transport and leads to plant architectural abnormality in Arabidopsis. Plant Cell, 2006,18(2):308-20.

[162] Xu J, Li Y, Wang Y, Liu H, Lei L, Yang H, Liu G, Ren D. Activation of MAPK kinase 9 induces ethylene and camalexin biosynthesis and enhances sensitivity to salt stress in Arabidopsis. J Biol Chem., 2008, 283(40):26996-7006.

[163] Liu XM, Kim KE, Kim KC, Nguyen XC, Han HJ, Jung MS, Kim HS, Kim SH, Park HC, Yun DJ, Chung WS. Cadmium activates Arabidopsis MPK3 and MPK6 via accumulation of reactive oxygen species. Phytochemistry, 2010, 71(5-6):614-8.

[164] Wurzinger B, Mair A, Pfister B, Teige M. Cross-talk of calcium-dependent protein kinase and MAP kinase signaling. Plant Signal Behav., 2011, 6(1):8-12.

[165] Jonak C, Hirt H. Glycogen synthase kinase 3/SHAGGY-like kinases in plants: an emerging family with novel functions. Trends Plant Sci., 2002, 7(10):457-61.

[166] Koh SH, Kim Y, Kim HY, Hwang S, Lee CH, Kim SH. Inhibition of glycogen synthase kinase-3 suppresses the onset of symptoms and disease progression of G93A-SOD1 mouse model of ALS. Exp Neurol., 2007, 205(2):336-46.

[167] Mahfouz MM, Kim S, Delauney AJ, Verma DP. Arabidopsis TARGET OF RAPAMYCIN interacts with RAPTOR, which regulates the activity of S6 kinase in response to osmotic stress signals. Plant Cell, 2006, 18(2):477-90.

[168] Santos MO, Aragao FJ. Role of SERK genes in plant environmental response. Plant Signal Behav., 2009, 4(12):1111-3.

[169] Boudsocq M, Barbier-Brygoo H, Lauriere C. Identification of nine sucrose nonfermenting 1-related protein kinases 2 activated by hyperosmotic and saline stresses in Arabidopsis thaliana. J Biol Chem., 2004, 279(40):41758-66.

[170] Umezawa T, Nakashima K, Miyakawa T, Kuromori T, Tanokura M, Shinozaki K, Yamaguchi-Shinozaki K. Molecular basis of the core regulatory network in ABA responses: sensing, signaling and transport. Plant Cell Physiology, 2010, 51(11):1821-39.

[171] Kulik A, Wawer I, Krzywińska E, Bucholc M, Dobrowolska G. SnRK2 protein kinases--key regulators of plant response to abiotic stresses. OMICS., 2011, 15(12):859-72.

[172] Shen Q, Gomez-Cadenas A, Zhang P, Walker-Simmons MK, Sheen J, Ho TH. Dissection of abscisic acid signal transduction pathways in barley aleurone layers. Plant Mol Biology, 2001, 47(3):437-48.

[173] Kobayashi Y, Murata M, Minami H, Yamamoto S, Kagaya Y, Hobo T, Yamamoto A, Hattori T. Abscisic acid-activated SnRK2 protein kinases function in the gene-regulation pathway of ABA signal transduction by phosphorylating ABA response element-binding factors. Plant Journal, 2005,44(6):939-49.

[174] Huai J, Wang M, He J, Zheng J, Dong Z, Lv H, Zhao J, Wang G. Cloning and characterization of the SnRK2 gene family from *Zea mays*. Plant Cell Reports, 2008, 27(12):1861-8.

[175] Kudla J, Batistic O, Hashimoto K. Calcium signals: the lead currency of plant information processing. *The* Plant Cell, 2010, 22(3):541-63.

[176] Geiger D, Scherzer S, Mumm P, Marten I, Ache P, Matschi S, Liese A, Wellmann C, Al-Rasheid KA, Grill E, Romeis T, Hedrich R. Guard cell anion channel SLAC1 is regulated by CDPK protein kinases with distinct $Ca^{2+}$ affinities. Proc Natl Acad Sci USA., 2010, 107(17):8023-8.

[177] Lee SJ, Cho DI, Kang JY, Kim MD, Kim SY. AtNEK6 interacts with ARIA and is involved in ABA response during seed germination. Molecules and Cells, 2010, 29(6):559-66.

[178] Kerk D, Bulgrien J, Smith DW, Barsam B, Veretnik S, Gribskov M. The complement of protein phosphatase catalytic subunits encoded in the genome of Arabidopsis. Plant Physiology, 2002, 129(2):908-25.

[179] Merlot S, Gosti F, Guerrier D, Vavasseur A, Giraudat J. The ABI1 and ABI2 protein phosphatases 2C act in a negative feedback regulatory loop of the abscisic acid signalling pathway. *The* Plant Journal, 2001, 25(3):295-303.

[180] Umezawa T, Sugiyama N, Mizoguchi M, Hayashi S, Myouga F, Yamaguchi-Shinozaki K, Ishihama Y, Hirayama T, Shinozaki K. Type 2C protein phosphatases directly regulate abscisic acid-activated protein kinases in Arabidopsis. Proc Natl Acad Sci USA., 2009, 106(41):17588-93.

[181] Hubbard KE, Nishimura N, Hitomi K, Getzoff ED, Schroeder JI. Early abscisic acid signal transduction mechanisms: newly discovered components and newly emerging questions. Genes & Dev*elopment,* 2010, 24(16):1695-708.

[182] Takemiya A, Kinoshita T, Asanuma M, Shimazaki K. Protein phosphatase 1 positively regulates stomatal opening in response to blue light in *Vicia faba*. Proc Natl Acad Sci USA., 2006, 103(36):13549-54.

[183] Yu RM, Zhou Y, Xu ZF, Chye ML, Kong RY. Two genes encoding protein phosphatase 2A catalytic subunits are differentially expressed in rice. Plant Molecular Biol*ogy,* 2003, 51(3):295-311.

[184] Yu RM, Wong MM, Jack RW, Kong RY. Structure, evolution and expression of a second subfamily of protein phosphatase 2A catalytic subunit genes in the rice plant (*Oryza sativa* L.). Planta, 2005, 222(5):757-68.

[185] Luan S. Protein phosphatases in plants. *Annual Review of Plant Biology,* 2003, 54:63-92.

[186] Fordham-Skelton AP, Chilley P, Lumbreras V, Reignoux S, Fenton TR, Dahm CC, Pages M, Gatehouse JA. A novel higher plant protein tyrosine phosphatase interacts with SNF1-related protein kinases via a KIS (kinase interaction sequence) domain. *The* Plant Journal, 2002, 29(6):705-15.

[187] Kerk D, Conley TR, Rodriguez FA, Tran HT, Nimick M, Muench DG, Moorhead GB. A chloroplast-localized dual-specificity protein phosphatase in Arabidopsis contains a phylogenetically dispersed and ancient carbohydrate-binding domain, which binds the polysaccharide starch. *The* Plant Journal, 2006, 46(3):400-13.

[188] Rushton PJ, Somssich IE. Transcriptional control of plant genes responsive to pathogens. Current Opinion in Plant Biol*ogy,* 1998, 1(4):311-5.

[189] Chaves MM, Oliveira MM. Mechanisms underlying plant resilience to water deficits: prospects for water-saving agriculture. Journal of Experimental Botany, 2004, 55(407):2365-84.

[190] Yi N, Kim YS, Jeong MH, Oh SJ, Jeong JS, Park SH, Jung H, Choi YD, Kim JK. Functional analysis of six drought-inducible promoters in transgenic rice plants throughout all stages of plant growth. Planta, 2010, 232(3):743-54.

[191] Jaleel CA, Gopi R, Sankar B, Gomathinayagam M, Panneerselvam R. Differential responses in water use efficiency in two varieties of Catharanthus roseus under drought stress. *Comptes Rendus Biologies,* 2008, 331(1):42-7.

[192] Verslues PE, Bray EA. Role of abscisic acid (ABA) and Arabidopsis thaliana ABA-insensitive loci in low water potential-induced ABA and proline accumulation. Journal of Experimental Botany, 2006, 57(1):201-12.

[193] Bray EA. Genes commonly regulated by water-deficit stress in Arabidopsis thaliana. Journal of Experimental Botany, 2004, 55(407):2331-41.

[194] Okamuro JK, Caster B, Villarroel R, Van Montagu M, Jofuku KD. The AP2 domain of APETALA2 defines a large new family of DNA binding proteins in Arabidopsis. Proc Natl Acad Sci USA., 1997, 94(13):7076-81.

[195] Zhou J, Tang X, Martin GB. The Pto kinase conferring resistance to tomato bacterial speck disease interacts with proteins that bind a cis-element of pathogenesis-related genes. EMBO J*ournal,* 1997, 16(11):3207-18.

[196] Ohme-Takagi M, Shinshi H. Ethylene-inducible DNA binding proteins that interact with an ethylene-responsive element. *The* Plant Cell, 1995, 7(2):173-82.

[197] Weigel D. The APETALA2 domain is related to a novel type of DNA binding domain. *The* Plant Cell, 1995, 7(4):388-9.

[198] Moose SP, Sisco PH. Glossy15, an APETALA2-like gene from maize that regulates leaf epidermal cell identity. Genes & Dev*elopment,* 1996, 10(23):3018-27.

[199] Liu Q, Kasuga M, Sakuma Y, Abe H, Miura S, Yamaguchi-Shinozaki K, Shinozaki K. Two transcription factors, DREB1 and DREB2, with an EREBP/AP2 DNA binding domain separate two cellular signal transduction pathways in drought- and low-temperature-responsive gene expression, respectively, in Arabidopsis. *The* Plant Cell, 1998, 10(8):1391-406.

[200] Agarwal PK, Agarwal P, Reddy MK, Sopory SK. Role of DREB transcription factors in abiotic and biotic stress tolerance in plants. Plant Cell Reports, 2006, 25(12):1263-74.

[201] Dubouzet JG, Sakuma Y, Ito Y, Kasuga M, Dubouzet EG, Miura S, Seki M, Shinozaki K, Yamaguchi-Shinozaki K. OsDREB genes in rice, Oryza sativa L., encode transcription activators that function in drought-, high-salt- and cold-responsive gene expression. *The* Plant Journal, 2003, 33(4):751-63.

[202] Abe H, Urao T, Ito T, Seki M, Shinozaki K, Yamaguchi-Shinozaki K. Arabidopsis AtMYC2 (bHLH) and AtMYB2 (MYB) function as transcriptional activators in abscisic acid signaling. *The* Plant Cell, 2003, 15(1):63-78.

[203] Liang YK, Dubos C, Dodd IC, Holroyd GH, Hetherington AM, Campbell MM. AtMYB61, an R2R3-MYB transcription factor controlling stomatal aperture in Arabidopsis thaliana. Current Biology, 2005, 15(13):1201-6.

[204] Rushton DL, Tripathi P, Rabara RC, Lin J, Ringler P, Boken AK, Langum TJ, Smidt L, Boomsma DD, Emme NJ, Chen X, Finer JJ, Shen QJ, Rushton PJ. WRKY transcription factors: key components in abscisic acid signalling. Plant Biotechnology *Journal,* 2012, 10(1):2-11.

[205] Hu H, Dai M, Yao J, Xiao B, Li X, Zhang Q, Xiong L. Overexpressing a NAM, ATAF, and CUC (NAC) transcription factor enhances drought resistance and salt tolerance in rice. Proc Natl Acad Sci USA., 2006, 103(35):12987-92.

[206] Xu K, Xu X, Fukao T, Canlas P, Maghirang-Rodriguez R, Heuer S, Ismail AM, Bailey-Serres J, Ronald PC, Mackill DJ. Sub1A is an ethylene-response-factor-like gene that confers submergence tolerance to rice. Nature, 2006, 442(7103):705-8.

[207] Bailey-Serres J, Voesenek LA. Flooding stress: acclimations and genetic diversity. *Annual Review of Plant Biology,* 2008, 59:313-39.

[208] Mustroph A, Lee SC, Oosumi T, Zanetti ME, Yang H, Ma K, Yaghoubi-Masihi A, Fukao T, Bailey-Serres J. Cross-kingdom comparison of transcriptomic adjustments to low-oxygen stress highlights conserved and plant-specific responses. Plant Physiology, 2010, 152(3):1484-500.

[209] Klok EJ, Wilson IW, Wilson D, Chapman SC, Ewing RM, Somerville SC, Peacock WJ, Dolferus R, Dennis ES. Expression profile analysis of the low-oxygen response in Arabidopsis root cultures. *The* Plant Cell, 2002, 14(10):2481-94.

[210] van Dongen JT, Frohlich A, Ramírez-Aguilar SJ, Schauer N, Fernie AR, Erban A, Kopka J, Clark J, Langer A, Geigenberger P. Transcript and metabolite profiling of the adaptive response to mild decreases in oxygen concentration in the roots of arabidopsis plants. Annals of Botany, 2009, 103(2):269-80.

[211] Branco-Price C, Kaiser KA, Jang CJ, Larive CK, Bailey-Serres J. Selective mRNA translation coordinates energetic and metabolic adjustments to cellular oxygen deprivation and reoxygenation in *Arabidopsis thaliana. The* Plant Journal, 2008, 56(5):743-55.

[212] Hoeren FU, Dolferus R, Wu Y, Peacock WJ, Dennis ES. Evidence for a role for AtMYB2 in the induction of the Arabidopsis alcohol dehydrogenase gene (ADH1) by low oxygen. Genetics, 1998, 149(2):479-90.

[213] Christianson JA, Wilson IW, Llewellyn DJ, Dennis ES. The low-oxygen-induced NAC domain transcription factor ANAC102 affects viability of Arabidopsis seeds following low-oxygen treatment. Plant Physiology, 2009, 149(4):1724-38.

[214] Licausi F, van Dongen JT, Giuntoli B, Novi G, Santaniello A, Geigenberger P, Perata P. HRE1 and HRE2, two hypoxia-inducible ethylene response factors, affect anaerobic responses in Arabidopsis thaliana. *The* Plant Journal, 2010, 62(2):302-15.

[215] Lasanthi-Kudahettige R, Magneschi L, Loreti E, Gonzali S, Licausi F, Novi G, Beretta O, Vitulli F, Alpi A, Perata P. Transcript profiling of the anoxic rice coleoptile. Plant Physiology, 2007, 144(1):218-31

[216] Rizhsky L, Liang H, Shuman J, Shulaev V, Davletova S, Mittler R. When defense pathways collide. The response of Arabidopsis to a combination of drought and heat stress. Plant Physiology, 2004, 134(4):1683-96

[217] Branco-Price C, Kawaguchi R, Ferreira RB, Bailey-Serres J. Genome-wide analysis of transcript abundance and translation in Arabidopsis seedlings subjected to oxygen deprivation. Annals of Botany, 2005, 96(4):647-60

[218] Davenport R, James RA, Zakrisson-Plogander A, Tester M, Munns R. Control of sodium transport in durum wheat. Plant Physiology, 2005, 137(3):807-18.

[219] Hasegawa PM, Bressan RA, Zhu JK, Bohnert HJ. Plant cellular and molecular responses to high salinity. Annu Rev Plant Physiol Plant Mol Biol., 2000, 51:463-499.

[220] Chinnusamy V, Zhu JK, Sunkar R. Gene regulation during cold stress acclimation in plants. Methods in Molecular Biology, 2010, 639:39-55.

[221] Munns R, Tester M. Mechanisms of salinity tolerance. *Annual Review of Plant Biology,* 2008, 59:651-81.

[222] Mahajan S, Pandey GK, Tuteja N. Calcium- and salt-stress signaling in plants: shedding light on SOS pathway. Arch Biochem Biophys., 2008, 471(2):146-58.

[223] Gaxiola RA, Li J, Undurraga S, Dang LM, Allen GJ, Alper SL, Fink GR. Drought- and salt-tolerant plants result from overexpression of the AVP1 $H^+$-pump. Proc Natl Acad Sci USA., 2001, 98(20):11444-9.

[224] Zhang HX, Blumwald E. Transgenic salt-tolerant tomato plants accumulate salt in foliage but not in fruit. Nature Biotechnology, 2001, 19(8):765-8.

[225] Xu D, Duan X, Wang B, Hong B, Ho T, Wu R. Expression of a Late Embryogenesis Abundant Protein Gene, HVA1, from Barley Confers Tolerance to Water Deficit and Salt Stress in Transgenic Rice. Plant Physiology, 1996, 110(1):249-257.

[226] Brini F, Hanin M, Lumbreras V, Amara I, Khoudi H, Hassairi A, Pagès M, Masmoudi K. Overexpression of wheat dehydrin DHN-5 enhances tolerance to salt and osmotic stress in Arabidopsis thaliana. Plant Cell Reports, 2007, 26(11):2017-26.

[227] Golldack D, Lüking I, Yang O. Plant tolerance to drought and salinity: stress regulating transcription factors and their functional significance in the cellular transcriptional network. Plant Cell Reports, 2011, 30(8):1383-91.

[228] Chinnusamy V, Zhu J, Zhu JK. Cold stress regulation of gene expression in plants. Trends Plant Sci., 2007, 12(10):444-51.

[229] Lee BH, Henderson DA, Zhu JK. The Arabidopsis cold-responsive transcriptome and its regulation by ICE1. *The* Plant Cell, 2005, 17(11):3155-75.

[230] Shinwari ZK, Nakashima K, Miura S, Kasuga M, Seki M, Yamaguchi-Shinozaki K, Shinozaki K. An Arabidopsis gene family encoding DRE/CRT binding proteins involved in low-temperature-responsive gene expression. Biochem Biophys Res Commun., 1998, 250(1):161-70.

[231] Medina J, Bargues M, Terol J, Pérez-Alonso M, Salinas J. The Arabidopsis CBF gene family is composed of three genes encoding AP2 domain-containing proteins whose expression Is regulated by low temperature but not by abscisic acid or dehydration. Plant Physiology, 1999, 119(2):463-70.

[232] Hannah MA, Wiese D, Freund S, Fiehn O, Heyer AG, Hincha DK. Natural genetic variation of freezing tolerance in Arabidopsis. Plant Physiology, 2006, 142(1):98-112.

[233] Chinnusamy V, Ohta M, Kanrar S, Lee BH, Hong X, Agarwal M, Zhu JK. ICE1: a regulator of cold-induced transcriptome and freezing tolerance in Arabidopsis. Genes & Dev*elopment*, 2003, 17(8):1043-54.

[234] Benedict C, Geisler M, Trygg J, Huner N, Hurry V. Consensus by democracy. Using meta-analyses of microarray and genomic data to model the cold acclimation signaling pathway in Arabidopsis. Plant Physiology, 2006, 141(4):1219-32.

Section 3

# Antimicrobial and Antioxidant Potential of Plant Extracts

Chapter 7

# *In vitro* Antioxidant Analysis and the DNA Damage Protective Activity of Leaf Extract of the *Excoecaria agallocha* Linn Mangrove Plant

C. Asha Poorna, M.S. Resmi and E.V. Soniya

Additional information is available at the end of the chapter

http://dx.doi.org/10.5772/55416

## 1. Introduction

Reactive oxygen species (ROS) are various forms of activated oxygen, which include free-radicals, e.g., superoxide anions ($O_2^-$), hydroxyl radicals ($\cdot OH$), non-free-radical compounds ($H_2O_2$) and singlet oxygen ($^1O_2$), which can be formed by different mechanisms in living organisms. Oxidative damage of DNA molecules associated with electron-transfer reactions is an important phenomenon in living cells, which can lead to mutations and contribute to carcinogenesis and the aging processes. ROS species are considered as important causative factors in the development of certain diseases such as diabetes, stroke, arteriosclerosis, cancer and cardiovascular diseases, in addition to the aging process. Prior administration of antioxidant provides a close relationship between FRSA and the involvement of endocrinological responses, which help to reverse the effect [1, 2]. Plants are rich sources of phytochemicals such as saponin, tannin, flavanoids, phenolic and alkaloids, which possess a variety of biological activities including antioxidant potential. Antioxidants provide protection to living organisms from damage caused by uncontrolled production of ROS and concomitant lipid peroxidation, protein damage and DNA stand breaking. Natural antioxidants are in high demand for application as bio-pharmaceuticals, nutraceuticals and food additives.

Terrestrial plants are considered potent sources of bioactive compounds and pharmacologically active compounds, however, little is known about the therapeutical potential of mangrove plants. Exploration of the chemical constituents of mangrove plants is necessary to find new therapeutic agents and this information is very important to the local community. Important reasons for studying the chemical constituents of mangrove plants are first, mangroves are a type of tropical forest that grows easily and has not as yet been widely utilized. Secondly, the chemical aspects of mangrove plants are very important because of the potential to develop compounds of agrochemical and medical value.

The plants of the genus *Excoecaria* (family: Euphorbiaceae) comprise nearly 40 species which are distributed throughout the mangrove regions of tropical Africa, Asia and northwest Australia. The most widely reported mangrove species is *Excoecaria agallocha* Linn. The latex of this plant has been used as a purgative and abortifacient, as well as in the treatment of ulcers, rheumatism, leprosy and paralysis. The leaves and latex of this tree have been used as fish poison in India, New Caledonia and Malaysia. The bark and wood is used in Thailand as a remedy for flatulence. Recently, much attention has been paid to *Excoecaria* species due to their cytotoxic and anti-HIV activities [3].

In this study we investigated the antimicrobial and antioxidant potential of methanol extract of *Excoecaria agallocha* Linn leaf. The DPPH and the oxidative DNA damage preventive activity and antioxidant potential of the crude methanol extract and sequential hexane, water and methanol extract of *E. agallocha* Linn leaf were also investigated. We found that water extract of *E. agallocha* Linn was more effective and could scavenge reactive oxygen species (ROS) thus preventing DNA strand scission by •OH generated in the Fenton reaction on pCAMBIA1301 DNA.

## 2. Material and methods

### 2.1. Plant materials and extraction procedure

The plants of *E. agallocha* were collected during the month of November 2009 from Ayiramthengu located near Alleppy in Kerala, India, at an average temperature of 28 - 34 °C. Collected plant material was dried in the shade and the leaves were then separated from the stem and pulverized to a fine powder in a grinder. The powdered leaf (10 g) was extracted sequentially with 100 ml of hexane and methanol by Soxhlet at a temperature not exceeding the boiling point of the solvent. To hexane extract water was added and separated in a separating funnel. The extracts were filtered using Whatman No. 1 filter paper and then concentrated in a vacuum at 40 °C using a rotary evaporator. The residues obtained were stored in a deep freezer at -80 °C until use.

## 3. Determination of antimicrobial activities

### 3.1. Microorganisms

Microbial cultures *Bacillus subtilis* (MTCC- 441), *S. pyogenes* (MTCC- 442), *E. coli* (MTCC - 443), *Salmonella typhi* (MTCC - 733), *K. pneumoniae* (MTCC - 109), *S. marcescens* (MTCC -*97), *Vibrio cholerae* (01) and *Vibrio cholerae* (08) belonging to bacterial species and *Candida albicans* (MTCC - 3017) yeast were used in this study. Microorganisms were provided by IMTech, Chandigarh, India.

### 3.2. Antimicrobial activity by Disc-diffusion assay

The dried plant extracts were dissolved in the same solvent (methanol and distilled water) to a final concentration of 100 mg/ml and sterilized by filtration by 0.45 μm Millipore filters.

Antimicrobial tests were then carried out by disc-diffusion method [4] using 100 µl of suspension containing $10^8$ CFU/ml of bacteria spread on nutrient agar (NA). The discs (6 mm in diameter) were impregnated with 5 µl of the extracts (500 µg/disc) at the concentration of 100 mg/ml and placed on the inoculated agar. Negative controls were prepared using the same solvents employed to dissolve the plant extracts. Ampicillin (10 µg/disc) was used as a positive reference standard to determine the sensitivity of one strain/isolate in each microbial species tested. The inoculated plates were incubated at 37 °C for 24 h for clinical bacterial strains. Antimicrobial activity was evaluated by measuring the zone of inhibition against the test organisms. Each assay in this experiment was repeated twice.

## 4. Evaluation of antioxidant activity

### 4.1. Reducing power assay

The reducing power of *E. agallocha* was determined as per the reported method [5]. Different concentrations of plant extract (100 -2000 µg/µl) in 1 ml of methanol were mixed with a phosphate buffer (2.5 ml, 0.2 M, pH 6.6) and potassium ferrocyanide (2.5 ml, 1 %). The mixture was incubated at 50 °C for 20 min. A portion (2.5 ml) of trichloroacetic acid (10 %) was added to the mixture, which was then centrifuged at 3000 rpm for 10 min. The upper layer of the solution (2.5 ml) was mixed with distilled water (2.5 ml) and $FeCl_3$ (0.5 ml, 0.1 %), and the absorbance was measured at 700 nm and compared with standards. Increased absorbance of the reaction mixture indicated increased reducing power.

### 4.2. Metal chelating effect

The chelation of ferrous ions by the extract was estimated as per the method of Dinis [6]. Different concentrations of the extract (100-2000 µg/µl) were added to a solution of 1 mM $FeCl_2$ (50 µl). The reaction was initiated by the addition of 1 mM ferrozine (0.1 ml) and the mixture was finally quantified to 1 ml with methanol, shaken vigorously and left standing at room temperature for 10 min. After the mixture had reached equilibrium, the absorbance of the solution was measured spectrophotometrically at 562 nm. All analyses were done in triplicate and average values were taken. The percentage of inhibition of ferrozine-$Fe^{2+}$ complex formation was calculated using the formula given below: % Inhibition $[(A_0 - A_1)/A_0 \times 100]$, where $A_0$ is the absorbance of the control and $A_1$ is the absorbance in the presence of the sample of *Excocaria* extract. $FeCl_2$ and ferrozine complex formation molecules are present in the control.

### 4.3. Nitric oxide radical inhibition activity

Nitric oxide, generated from sodium nitroprusside in an aqueous solution at physiological pH, interacts with oxygen to produce nitrite ions which were measured by Griess reaction [7]. The reaction mixture (3 ml) containing sodium nitroprusside (10 mM) in a phosphate buffer saline and the extract (100 – 2000 µg/µl) were incubated at 25°C for 150 min. After incubation, 0.5 ml of the reaction mixture was removed and 0.5 ml of Griess reagent (1 % (w/v) sulfanilamide, 2 % (v/v) $H_3PO_4$ and 0.1 % (w/v) naphthylethylene diamine hydrochloride were added. The absorbance of the chromophore formed was measured at 546 nm.

## 4.4. Lipid peroxidation and thiobarbituric acid reaction

A modified TBARS assay [8] was used to measure the lipid peroxide formed using egg yolk homogenate as lipid rich media [9]. Egg homogenate (0.5 ml of 10 %, v/v) and 0.1 ml of extract were added to a test tube and made up to 1 ml with distilled water, 0.05 ml of $FeSO_4$ (0.07 M) was added to induce lipid peroxidation and the mixture was incubated for 30 min. Then, 1.5 ml of 20 % acetic acid (pH 3.5) and 1.5 ml of 0.8 % (w/v) thiobarbituric acid in 1.1 % sodium dodecyl sulphate were added, the resulting mixture was vortexed and then heated at 95 °C for 1 h. After cooling, 5.0 ml of butanol was added to each tube and centrifuged at 3000 rpm for 10 min. The absorbance of the organic upper layer was measured at 532 nm. Inhibition of lipid peroxidation percent by the extract was calculated as 100- $[(A_1/A_2) \times 100]$; where $A_1$ is the absorbance value in the presence of extract and $A_2$ of the fully oxidized control.

## 4.5. Determination of DPPH radical scavenging capacity

Quantitative estimation of the free-radical scavenging activity was measured by DPPH assay [10]. The reaction mixture contained a different concentration (100-2000μg/μl) of test extract and 2.9 ml of DPPH (60 μM) in methanol. These reaction mixtures were taken in test tubes and incubated at 37 °C for 30 min, the absorbance was measured at 517 nm. The percentage of radical scavenging activity by the sample treatment was determined by comparison with the methanol treated control group. BHT and ascorbic acid was used as a positive control. The DPPH radical concentration was calculated using the following equation: scavenging effect (%) = $(DPPH^{\cdot})_T$ / $(DPPH^{\cdot})_{T=0}$ x 100, where $(DPPH^{\cdot})_T$ is the concentration of $DPPH^{\cdot}$ at 30 min time and $(DPPH^{\cdot})_{T=0}$, the concentration at zero time (initial concentration).

## 4.6. Total antioxidant activity

The assay is based on the reduction of Mo (VI) to Mo (V) by the extract and subsequent formation of a green phosphate / Mo (V) complex at the acid pH [11]. The tubes containing 0.1 ml of the extract and the 1 ml of reagent solution (0.6 M sulphuric acid, 28 mM sodium phosphate and 4 mM Ammonium molybedate) were incubated at 95 °C for 90 min. After the mixture was cooled to room temperature, absorbances were taken at 695 nm against the blank. The antioxidant capacity was expressed as AAE.

## 4.7. DNA nicking induced by hydroxyl radical

The DNA damage protective activity of *E. agallocha* L. extract was performed using super coiled pCAMBIA1301 DNA. Plasmid DNA isolation was done using GenElute™ Plasmid

Miniprep Kit (Sigma - Aldrich USA). A mixture of 10 µl of hexane, water and methanol extract (100 µg/µl), and plasmid DNA (0.5µg) was incubated for 10 min at room temperature followed by the addition of 10 µl of Fenton's reagent (30 mM $H_2O_2$, 50 µM ascorbic acid and 80 µM of $FeCl_3$). The final volume of the mixture was made up to 20 µl with double distilled water and incubated for 30 min at 37 °C. The DNA was analysed on 1 % agarose gel using ethidium bromide staining and photographed in Gel Doc. Quercetin (50 µM) was used as a positive control [12].

### 4.8. Phytochemical analysis

Chemical tests were carried out on the aqueous extract of the powdered specimens using standard procedures to identify the constituents as described by Harborne [13].

## 5. Results and discussion

### 5.1. Antimicrobial activity

This paper illustrates the antimicrobial, antioxidant and DNA protective effect of *Excoecaria agallocha* Linn leaf extract. The phytochemical constituent of *E. agallocha* Linn can be summarized as follows: methanolic leaf extract shows the presence of saponin, tannin and a high content of terpinoids, whereas cardiac glycosides are absent. The antimicrobial activities of methanol extract of *Excoecaria agallocha* Linn leaf is summarized in Table 1. The extract of *E. agallocha* reported to have significant *in vitro* antibacterial activity against *Staphylococcus aureus, Shigella dysenteriae, Shigella sonnei* and *Enterococci* with the zones of inhibition ranging from 11 to 15 mm, and no significant activity was reported against *Shigella flexneri* and *Staphylococcus epidermis* at test concentrations [14].

| Microbes / Sample | +ve | 1 | 2 | 3 | 4 |
|---|---|---|---|---|---|
| *B.subtilis* | 17 | NI | NI | NI | NI |
| *S.pyogenes* | NI | NI | NI | NI | NI |
| *E.coli* | 13 | 12 | NI | NI | 10 |
| *S.typhi* | 15 | 12 | NI | NI | 11 |
| *K.pneumoniae* | >10 | 12 | NI | NI | 10 |
| *S.marcescens* | 36 | 11 | 10 | NI | 10 |
| *V. Cholerae (01)* | 10 | 11 | 10 | NI | 10 |
| *V. Cholerae (08)* | 15 | 10 | NI | NI | 10 |

Ampicillin used as positive control and the solvent used as negative control, activity expressed in millimole – mm concentration. (NI- no inhibition).

**Table 1.** Antimicrobial activity of different fractions (500mg/6mm disc) of *Excoecaria agallocha* Linn.

## 5.2. Reducing power assay

The reducing power of *E. agallocha* and reference compound ascorbic acid increases steadily with the increase in concentration. The absorbances at 700 nm of *E. agallocha* 4.00 (2000µg/µl) and ascorbic acid 4.5 (1000 µg/µl) shows that *E. agallocha* can act as electron donor and can react with free-radicals to convert them to more stable products and thereby terminate radical chain reactions (Figure 1A). The $IC_{50}$ value of *E. agallocha* is observed to be at 62.96 µg. The reducing power of plant compounds might be due to the di- and mono-hydroxyl substitution in the aromatic ring which possesses potential hydrogen donating abilities [15]. The reducing properties are generally associated with the presence of reductones [16], which have been shown to exert antioxidant activity by breaking the free-radical chain by donating a hydrogen atom.

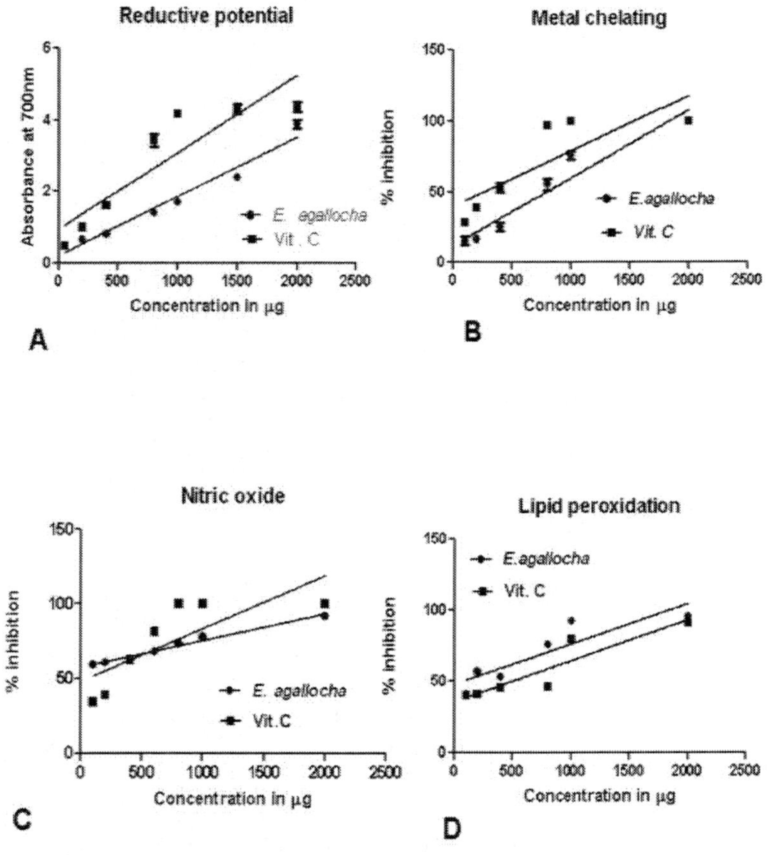


**Figure 1.** The concentration dependent (100 -2000 µg/µl) antioxidant activity: A) Reducing power, B) Metal chelating, C) Nitric oxide and D) Lipid peroxidation of methanol extract of leaf *E. agallocha* Linn (mean ± SD, $n = 3$).

### 5.3. Metal chelating effect

It has been proposed that transition metals catalyse the formation of the first few radicals to start the propagation of radical chain reaction in lipid peroxidation. Chelating agents may inhibit lipid oxidation by stabilizing transition metals. Ferrozine can quantitatively form complexes with $Fe^{2+}$. In the presence of other chelating agents, the complex formation is disrupted with the result that the red colour of the complex is decreased. As shown in Figure 1B, the ferrozine - $Fe^{2+}$ complex is not complete in presence of the plant extract, indicating its ability to chelate the iron. The absorbance of ferrozine-$Fe^{2+}$ complex decreased linearly in a dose dependent manner (100– 2000μg/μl) and the $IC_{50}$ value is estimated as 2.47 μg. The metal chelating activity of *E. agallocha* was evaluated against $Fe^{2+}$. The standard compounds ascorbic acid and BHT did not exhibit any metal chelating activity at the tested concentrations (100–2000 μg/μl). Reaction of ascorbic acid and gallic acid with $FeCl_2$ might enhance the degradation of ascorbic acid and gallic acid, and increase the ascorbyl and gallic acid radical concentrations [17].

### 5.4. Nitric oxide radical inhibition activity

The antioxidant system protects the pathogens against the ROS-induced oxidative damage. Nitric oxide radical generated from the sodium nitropruside is measured by the Greiss reduction. Sodium nitropruside at physiological pH spontaneously generates nitric oxide, which thereby interacts with oxygen to produce nitrate ions that can be estimated using Greiss reagents. Thus, the scavengers of nitric oxide compete with the oxygen, leading to reduced production of nitric oxide. The chromophore formed during diazotization of nitrite with sulphanilamide and its subsequent coupling with naphthyl ethylene diamine was read at 546 nm. The methanolic leaf extract of *E. agallocha* has shown a more significant effect than that of ascorbic acid and the results are explained in Figure 1C. The $IC_{50}$ of *E. agallocha* is estimated as 4.8 μg/μl.

### 5.5. Lipid peroxidation and thiobarbituric acid reaction

Egg yolk lipids undergo rapid non-enzymatic peroxidation when incubated in the presence of ferrous sulphate with subsequent formation of malonodialdehyde (MDA) and other aldehydes that form pink chromogen with TBA absorbing at 532 nm [18]. Peroxidation of lipids has been shown to be the cumulative effect of reactive oxygen species, which disturb the assembly of the membrane causing changes in fluidity and permeability, alterations of ion transport and inhibition of metabolic processes [19]. The extract of *E. agallocha* exhibited strong lipid peroxidation inhibition in a concentration dependent manner (Figure 1D). The $IC_{50}$ value for the inhibition of lipid peroxidation is observed to be 100 μg/μl. This activity was higher than that of ethanolic and hexane extract of *Ziziphus mauratiana* and *Z. spina-christi* reported using egg yolk as media of peroxidation [20]. The studies made on *E. agallocha* leaf extract suggest that it could play a role in protecting the physicochemical properties of membrane bilayers from free-radical induced severe cellular dysfunction.

## 5.6. Determination of DPPH radical scavenging capacity

Methanolic leaf extract of *E. agallocha* has shown good free-radical scavenging activity at all tested concentrations. DPPH· is a stable free-radical and can accept an electron or hydrogen radical becoming a stable diamagnetic molecule [21]. DPPH is purple in colour which turns yellow; the intensity of the yellow colour depends upon the amount and nature of radical scavenger present in the sample and standard compounds. The scavenging activity increases with an increase in concentration of the extract, as well as ascorbic acid, and levels off with further increases in concentration - $IC_{50}$ value is at 67.50 µg/µl (Figure 2A). The residual concentration of DPPH depends exclusively on the structure of the phenolic compound, since there are two theoretical termination reactions: one between DPPH radicals and the other between DPPH· and phenol radical (Phe O·). However, the former reaction cannot occur due to steric hindrance and the latter reaction competes with the Phe O· coupling termination reaction [22, 23]. The accessibility of the radical centre of DPPH· to each polyphenol could also influence the order of the antioxidant power. Recently, the free-radical scavenging potential possessed by *Desmodium gangeticum* chloroform root extract was reported [24].

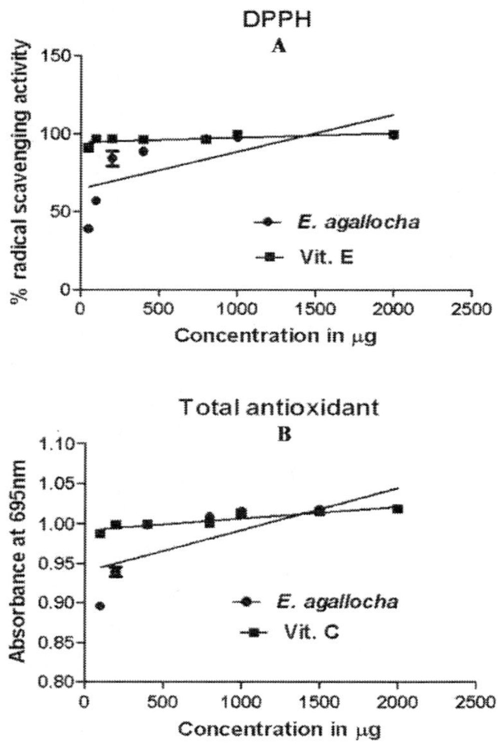


**Figure 2.** A) DPPH scavenging activity in concentration dependent (100 -2000 µg/µl) manner of methanol extract of leaf of *E. agallocha* Linn and B) Total antioxidant activity checked by phosphomolybdenum assay (mean ± SD, *n* = 3).

## 5.7. Total antioxidant activity

The total antioxidant potential of *E. agallocha* was investigated and compared against ascorbic acid, the results are explained in Figure 2B. The methanolic extract of *E. agallocha* is observed to be more effective - the $IC_{50}$ value is calculated as 3.36 µg. The phosphomolybdenum method usually detects antioxidants such as ascorbic acid, some phenolics, α-tocopherol and carotenoids [25]. Ascorbic acid, glutathione, cysteine, tocopherols, polyphenols and aromatic amines have the ability to donate hydrogen and electrons, and can thus be detected by this assay. Oxidative stress is the condition in which an imbalance between oxidant stimuli and physiological antioxidants exist leading to the damage of a cell. The body's physiological response to oxidative stress is through several antioxidant systems which include enzymes like superoxide dismutase, catalase, glutathione peroxidase and a variety of large molecules such as albumin and ferrtin, and small molecules such as ascorbic acid, tocopherol, etc. These antioxidants can be found as water-soluble or lipid-soluble molecules and localized transiently throughout tissues, cells and cell types.

## 5.8. DNA nicking induced by hydroxyl radical

Hydroxyl radical is the most reactive among reactive oxygen species, it has the shortest half life compared with others and is considered to be responsible for much of the biological damage in free-radical pathology. The radical has the capacity to cause strand breakage in DNA, which contributes to carcinogenesis, mutagenesis and cytotoxicity [11]. The DNA protective effect of hexane water and methanol extract of *E. agallocha* Linn leaf was checked against Fenton's induced DNA damage of pCAMBIA 1301 DNA. The protection offered against DNA damage by *E. agallocha* (10- 200 µg/µl) (results not shown) was concentration dependent. At concentration 100 µg/µl protection was more effective and slightly close to that of 5U of Catalase and 50 µM of quercetin tried (Figure 3). Native DNA has shown three forms, form I open circular form, form II single supercoiled band and below it form III, whereas DNA + Fenton's reagent and hexane fraction has exhibited complete degradation of DNA. Water fraction of *E. agallocha* Linn has shown very good protective activity and has retained all three forms. These results indicate that the water extract of *E. agallocha* Linn effectively mitigates the oxidative stresses on susceptible biomolecules, such as DNA.

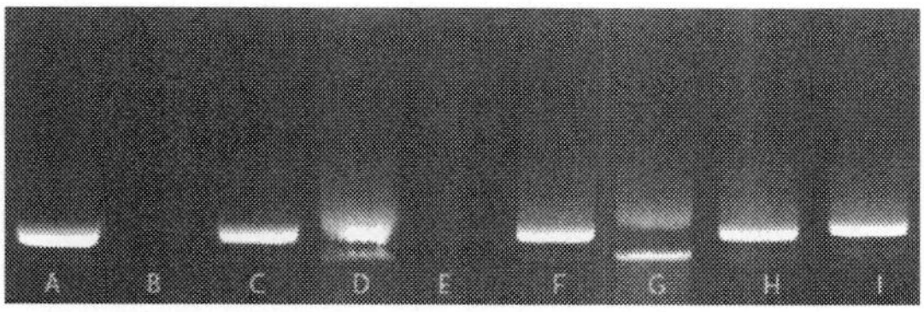


**Figure 3.** DNA protective effect of hexane, water and methanol extract of leaf of *E. agallocha* Linn, was checked against Fenton's induced DNA damage of pCAMBIA 1301 DNA. Effect of fraction of *E. agallocha* Linn. on oxidative DNA nicking caused by hydroxyl radicals. Lane A: native pCAMBIA 1301 DNA; Lane B: DNA + Fenton's reagent; Lane C: DNA + Fenton's reagent + *E. agallocha.* Linn water extract (50 µg/ml), Lane D: DNA + Fenton's reagent + *E. agallocha.* Linn methanol extract (100 µg/ml),

Lane E: DNA + Fenton's reagent + *E. agallocha.* Linn hexane extract, Lane F: DNA + Fenton's reagent + *E. agallocha.* Linn water extract (100μg/ml), Lane G: DNA + Fenton's reagent + *E. agallocha* Linn. crude methanol extract (100μg/ml) and Lane H: DNA + Fenton's reagent + quercetin (50 μM), Lane I: DNA + Fenton's reagent + Catalase (5 units).

## 6. Conclusion

In conclusion, the result obtained in the present study shows that the methanolic extract of *E. agallocha* contains a number of antioxidant compounds that can effectively scavenge ROS. Antioxidant properties of botanical extracts should be assessed in an array of model systems using several different indices because the effectiveness of such antioxidant material is largely dependent upon the chemical and physical properties of the system to which they are added and a single analytical protocol adopted to monitor lipid oxidation may not be sufficient to make a valid judgement. Hence, it may be concluded that the strong radical scavenging activity and oxidative DNA damage preventive activity of *E. agallocha* Linn may be correlated with its rich content of flavonoids.

## Author details

C. Asha Poorna, M.S. Resmi and E.V. Soniya*
*Plant Molecular Biology, Rajiv Gandhi Centre for Biotechnology, Government of India, Trivandrum, India*

## 7. References

[1] Alma, M.H., Mavi, A., Yildirim, A., Digrak, M., Hirata, T. Screening chemical composition and antioxidant and antimicrobial activities of the essential oils from *Origanum syriacum* L. growing in Turkey. *Biol. Pharm. Bull.* 2003, 26, pp 1725–1729.

[2] Verma A.R., Vijayakumar A, Rao C.V., Mathela C.S. In vitro and in vivo antioxidant properties and DNA damage protective activity of green fruit of *Ficus glomerata, Food and Chemical Toxicology*. 2010, 48, pp 704–709.

[3] Yin, B.W., Shen, L.R., Zhang, M.L., Zhao, L., Wang, Y.L., Huo, C.H., Shi, Q.W. Chemical Constituents of Plants from the Genus *Excoecaria. Chemistry and Biodiversity* 2008, 5, pp 2356- 2371.

[4] Murray, P.R., Baron, E.J., Pfaller, M.A., Tenover, F,C., Yolke, R,H. Manual of Clinical Microbiology, 6th ed. ASM, Washington, DC. Service, R.F, Antibiotics that resist resistance. *Science*. 1995, 270, pp 724–727.

[5] Brand, W.W., Cuvelier, M.E., Berset, C. Use of free radical method to evaluate antioxidant activity. *Lebensmittel- Wissenschaft und Technologie.* 1995, 28, pp. 25-30.

[6] Oyaizu, M. 1986, Studies on the products of browning reaction prepared from glucose amine. *Japan Journal of Nutrition*, 1986, 44. pp. 307– 315.

* Corresponding Author

[7] Marcocci, I., Maguire, J.J., Droy-Lefai, M.T., Parker, L. Antioxidant action of ginkgo biloba extracts EGb761. *Biochem. Biosphys. Res. Commun.* 1994, 201, pp. 748.

[8] Ohkawa, H., Ohishi, N., Yagi, K. Assay for lipid peroxides in animal tissues by thiobarbituric acid reaction. *Analytical Biochemistry*, 1979, 95, pp. 351–358.

[9] Ruberto, G., Baratta, M.T., Deans, S.G., Dorman, H,J,D. Antioxidant and antimicrobial activity of *Foeniculum vulgare* and *Crithmum maritimum* essential oils. *Planta. Med.* 2000, 66, pp. 687-693.

[10] Dinis, T.C.P., Madeira, V.M.C., Almeida, L.M. Action of phenolic derivates (acetoaminophen, salicylate, and 5-aminosalicylate) as inhibitors of membrane lipid peroxidation and as peroxyl radical scavengers. *Archive of Biochemistry and Biophysics.* 1994, 315, pp. 161–169.

[11] Prieto, P., Pineda, M., Aguilar, M. Spectrophotometric quantitation of antioxidant capacity through the formation of phosphomolybdenum complex: specific application to determination of vitamin E, *Analytical Biochemistry*, 1999, 269, pp. 337–341.

[12] Harbone, J.B. Phytochemical methods: *A guide to modern techniques of plant analysis*, 3rd Edition. Chapman and Hill, London.1998.

[13] Soares, J.R., Dins, T.C.P., Cunha, A,P., Almeida, L.M. Antioxidant activity of some extracts of *Thymus zygis*. *Free Radic. Res.*, 1997, 26, pp. 469-478.

[14] Poorna, C.A., Sathish K.M., Santhoshkumar T,R., Soniya E,V. Phytochemical analysis and *in vitro* screening for biological activities of *Acanthus ilicifolius*. *Journal of Pharmacy Research*, 2011, 4, (7), 1977-1981.

[15] Mathew, S., Abraham, T.E. *In vitro* antioxidant activity and scavenging effects of *Cinnamomum verum* leaf extract assayed by different methodologies, *Food and Chemical Toxicology*, 2006, 44, pp. 198–206.

[16] Subhan, N., Ashraful, M.A., Ahmed, F., Shahid, I,J., Nahar, L., Sarker, S.D. Bioactivity of *Excoecaria agallocha, Rev. Bras. De- Farmacogn.* 2008, 18, 4, pp. 521-526.

[17] Shimada, K., Fujikawa, K., Yahara, K., Nakamura, T. Antioxidative properties of xanthan on the auto oxidation of soybean oil in cyclodextrin emulsion. *Journal of Agricultural and Food Chemistry*, 1992, 40, pp. 945–948.

[18] Pin-Der-Duh, X. Antioxidant activity of burdock (*Arctium lappa* Linne): its scavenging effect on free radical and active oxygen. *Journal of the American Oil Chemists Society*, 1998, 75, pp. 455-461.

[19] Kosugi, H., Kato, T., Kikugawa, K. 1987. Formation of yellow, orange and red pigments in the reaction of alk-2-enals with 2-thiobarbituric acid. *Anal. Biochem.* 1987, 165. pp. 456-464.

[20] Abalaka, M.E., Mann, A., Adeyemo, S. O. Studies on in-vitro antioxidant and free radical scavenging potential and phytochemical screening of leaves of Ziziphus mauritiana L. and Ziziphus spinachristi L. compared with Ascorbic acid. J. Medi. Genet. Geno. 2011, 3(2), pp. 28–34.

[21] Nigam, S., Schewe, T. Phopholipase A2s and lipid peroxidation. *Biochim Biophy Acta.* 2000, 1488, pp. 167-181.

[22] Satoh, K., Sakagami, H. Effect of metal ions on radical intensity and cytotoxic activity of ascorbate. *Anticancer Research*, 1997, 17, pp. 1125-1130.

[23] Mahinda, S., Soo-Hyun, K., Nalin, S., Jin- Hwan, L., You-jin, J. Antioxidant potential of *Ecklonia cavaon* reactive oxygen species scavenging, metal chelating, reducing power and lipid peroxidation inhibition. *Food Sci. Technol Int.* 2006, 12, pp. 27-38.

[24] Srivats, S., Ramakrishnan, G., Paddikkala, J., Kurian, G. A. An in vivo and in vitro analysis of free radical scavenging potential possessed by Desmodium gangeticum chloroform root extract: Interpretation by gsms. *Pak. J. Pharm. Sci.* 2012, 25 (1), pp. 27-34.

[25] Hochestein, P., Atallah, A.S. The nature of antioxidant systems in the inhibition of mutation and cancer. *Mutat. Res.* 1988, 202, pp. 363-375.

Section 4

# Pollutants Analysis and Effects

Chapter 8

# Determination of Triazole Fungicides in Fruits and Vegetables by Liquid Chromatography-Mass Spectrometry (LC/MS)

Nor Haslinda Hanim Bt Khalil and Tan Guan Huat

Additional information is available at the end of the chapter

http://dx.doi.org/10.5772/55306

## 1. Introduction

Triazole pesticides derivatives represent the most important category of fungicides that have excellent protective, curative and eradicant power towards a wide spectrum of crop diseases [1]. The fungicide group, demethylation inhibitors (DMI), which contain the triazole fungicides, was introduced in the mid-1970s. These fungicides are highly effective against many different fungal diseases, especially powdery mildews, rusts, and many leaf-spotting fungi. [2].

The number of pesticides registered for use increases every year and many pesticides that have been banned for health reasons are also still being used illegally. And introduction of new pesticides in the field of residue analysis also cause the laboratories involved in the analysis to face more challenging task. This leads to the development of many multi-residue methods by various researchers [3-7].

In the past, pesticides and their degradation products, which are generally thermolabile, non-volatile and exhibit medium to high polarity have been analysed using GC with specific detectors such as ECD, NPD and FPD [8-12]. Due to the drawbacks of the separation techniques such as sensitivity, insufficient number of analytes that can be analysed and the need for confirmation either with different column polarity or detectors, GC/MS has become the primary approach to analyse all classes of GC-amenable pesticides [3, 13-14]. Later, HPLC combined with a diode array UV detector was established as a complementary technique to GC to analyse pesticides and their degradation products [15]. However it is not sufficient to use only the UV spectrum for identification of the analytes. Robust atmospheric pressure ionization (API) ion source designs, which consist of electrospray ionization (ESI) and atmospheric pressure chemical ionization (APCI) were developed and very powerful and reliable LC/MS instruments have been introduced commercially. The atmospheric pressure interfaces has been used to broaden the range of analytes to be analysed by liquid chromatography coupled with mass spectrometry [16-17].

Solvents such as acetone, ethyl acetate and acetonitrile may be used for extraction. However, acetonitrile is the recommended solvent and is being used widely for QuEChERS method because when salts are added, it separates more easily from water than acetone. Ethyl acetate has the advantage of partial miscibility with water but it co-extracts with lipids and waxes giving lower recoveries for acid-base pesticides [6]. A study by Lehotay S.J et al in 2010 [18] showed that results using acetate-buffered MeCN gave more accurate (true and precise) results for all analytes in LC-MS/MS than EtOAc. On the contrary EtOAc is a better solvent for GC rather than MeCN as demonstrated by the slightly more consistent recoveries and reproducibility overall in GC-MS using EtOAc.

## 2. Materials and methods

### 2.1. Equipment

i. LC/MS instrument

The chromatographic system used to analyse the extract is a Waters Alliance Separations Module 2695 equipped with a quaternary solvent delivery system, autosampler and column heater. A Waters ZQ 4000 single quadrupole Mass Spectrometer was used.

ii. Chopper and Vortex mixer

Robot Coupe R5 V.V (Jackson, MS) and OMNI mixer homogenizer (OMNI International, USA) were used to cut the fruit and vegetable samples into smaller pieces. Genie II vortex mixer was used to swirl the tubes.

iii. Centrifuge

Sorvall Legend RT Plus / Thermo Scientific were used for the centrifugation.

iv. Balance

A Shimadzu top-loading balance Libror AEG-220 was used to weigh the chopped samples and solid reagents and a Shimadzu analytical balance Libror EB-3200 HU was used in the preparation of stock standard solutions.

v. Vials and tubes

For the extraction step, 50mL centrifuge tubes were employed. 15mL graduated centrifuge tubes were used for dispersive SPE in the method.

vi. Solvent Evaporator

Zymark nitrogen evaporator Turbovap LP was used to concentrate the extracts and to facilitate solvent exchange when necessary.

## 2.2. Chemicals and reagents

The fungicides: cyproconazole, difenoconazole, fenbuconazole, hexaconazole, myclobutanil, propiconazole, tebuconazole, triadimefon and triadimenol were purchased from Pestanal, Riedel-de Haen (Seelze, Germany) with purity ranging from 95-100%. Acetonitrile, methanol, ethyl acetate of HPLC grade and residue analysis grade were obtained from Labscan and Merck (Darmstadt, Germany). Formic acid which was added to the mobile-phase acetonitrile was purchased from Fluka.

Salts used for the dispersive clean-up were anhydrous magnesium sulfate and sodium acetate which were obtained from Merck and Mallinckrodt. The SPE sorbent used was Bondesil PSA, 40μm from Varian. Deionized water (<8cm MΩ resistivity) was obtained from the Milli-Q Advantage A10 Pure Water System (Millipore, Bedford, MA, USA). All solvents were filtered using a 0.45μm nylon membrane filter from Whatman (Maidstone, England).

### *2.2.1. Stock and working solutions*

Stock solutions of 1000 μg/ml were prepared in methanol by dissolving approximately 0.020g of the individual standards in 20mL of methanol and stored at 4°C in a reagent bottle. Intermediate standard solution mixtures of 50 and 10 μg/ml were prepared in methanol and standard working solutions at various concentrations were prepared daily by appropriate dilution of the stock solution or the intermediate standard solution in methanol.

## 2.3. Methods

### *2.3.1. Extraction and clean-up*

The extraction method used was based on QuEChERS method [6] and modified by Aysal et al., 2007 [7]. The samples were chopped into smaller pieces and homogenised using a food processor. 30g of the homogenised sample was placed in a 250ml borosilicate bottle and extracted with 60ml of ethyl acetate, 30g of anhydrous sodium sulfate and 5g of sodium hydrogen carbonate. 10 ml of the extract was centrifuged at 2500 rpm for 2 min followed by clean-up with PSA sorbent and anhydrous magnesium sulfate. After clean-up, 5 mL of the extract was reduced to almost dryness under a stream of nitrogen and was redissolved in methanol.

### *2.3.2. Recovery studies*

Four types of fruits and vegetables namely carrot, cabbage, tomato and orange were used for the recovery studies which represent root and tuber vegetables, brassica leafy vegetables, fruiting vegetables and citrus fruits according to the CODEX classification of commodities.

The samples for recovery determination were prepared by spiking with the standard solution. Each sample was fortified with nine triazole standards at 0.05, 0.5 and 1.0 μg/ml and five replicates at these fortification levels for each matrix. The fortified samples were allowed to stand for 30 min before extraction to allow the spiked solutions to penetrate the samples and attain the fungicide distribution in the samples.

### *2.3.3. Calibration*

Quantification of triazoles were performed and compared by using calibration standards involving both matrix-matching by adding standards to blank extracts and non-matrix matching (standards in solutions) based on a calibration curve. For matrix matching, blank extracts were fortified with the pesticide working standard after dispersive clean-up. The calibration solutions were prepared daily at 7 levels of concentrations ranging from 0.05 to 2.0 μg/ml. The LOD's and LOQ's were calculated by multiplying the standard deviation of the calculated amount for each triazole by 3 and 10 respectively.

# 3. Results and discussion

## 3.1. High Performance liquid chromatography-mass spectrometry

### *3.1.1. HPLC*

A C18 reversed phase column (4.6mm x 75mm, 3.5 um particle size) was used in this study to generate less back pressure as it allows more flexibility to adjust the flow-rate. A short column was also used to obtain shorter separation times that produce narrower peaks because there is less time for diffusive broadening. The small particle size used helps to generate more pressure and generally give higher separation efficiencies. Smaller particle size column is necessary to maintain resolution in the short column used. The HPLC column had been run at different flow rates; 0.8 mL/min, 1 mL/min, 1.2 mL/min and 1.4 mL/min during optimization and it was found that it gives better resolution at a flow rate of 1.2 mL/min A common operating temperature is 40°C as higher temperature is better in producing sharp peaks and earlier elution [19]. For this study, the effects of column temperature were also evaluated at various temperatures; 20°C, 25°C, 30°C, 35°C and 40°C. Figure 1 showed that 25°C column temperature found to give better separation after running triazole standard mixture.

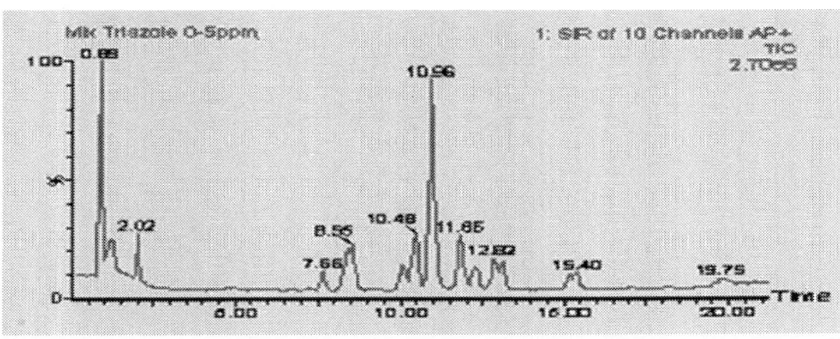


**Figure 1.** Acetonitrile/ $H_2O$ mobile phase, column temperature 25°C

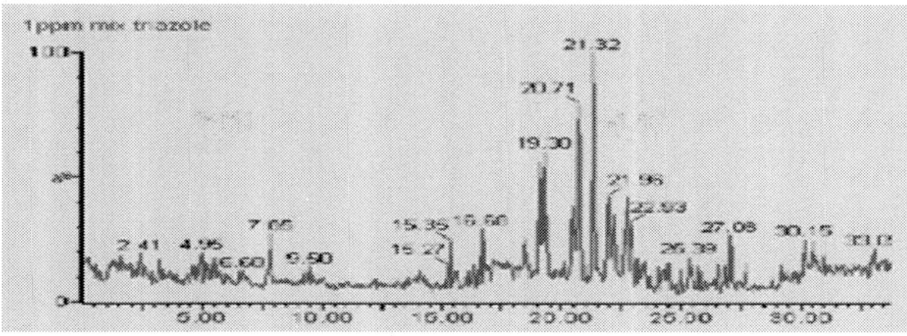


**Figure 2.** Acetonitrile/ $H_2O$ mobile phase, column temperature 35°C

In this work, water and acetonitrile with 0.1% acetic acid were used for all liquid chromatographic separations. No buffers were used. Two types of additives have been added to the mobile phase that is acetic acid and formic acid during optimization and it was found that by adding 0.1% of acetic acid gives better resolution than formic acid. The results are also more stable. The additive was added to improve the chromatographic shape and to provide a source of protons in the reversed phase and to enhance and control the formation of ions. A study on water:methanol with both 0.1% acetic acid and formic acid was also done but it did not give good resolution and the results are not reproducible.

The reversed phase solvents are installed on the channels A and C. Channel A is the aqueous solvent (water) and channel C is the organic solvent (acetonitrile). Silica dissolves at high pH, therefore it is not recommended to use solvents that exceed pH7. The pH for acetonitrile was in the range of pH 2.5 – pH 3.5.

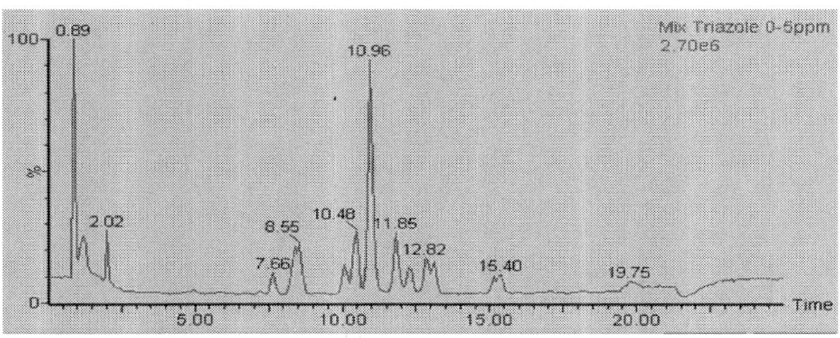


**Figure 3.** Chromatogram of triazole standards mixture at 0.5 μg/mL

## 3.2. Mass spectrometry

Prior to triazole analysis, the chromatographic parameters including the heated nebuliser parameters were optimized. LCMS infusion was carried out to examine the ionization and fragmentation patterns of the analytes. The APCI source was used in the positive ion mode. A full scan was used for the MS optimization and a selected ion monitoring (SIM) was used for the monitoring of the selected ion. Table 1 showed the triazoles and quantitation ion.

| Analyte | tR, min | Quantitation ion, m/z |
|---|---|---|
| Triadimenol | 7.66 | 296.1 |
| Cyproconazole | 8.55 | 292.1 |
| Myclobutanil | 10.08 | 289.1 |
| Triadimefon | 10.48 | 294.1 |
| Tebuconazole | 10.96 | 308.1 |
| Hexaconazole | 11.85 | 314.1 |
| Fenbuconazole | 12.25 | 337.1 |
| Propiconazole | 12.82 | 342.1 |
| Difenoconazole | 13.14 | 406.2 |

tR- Retention time

**Table 1.** List of Triazole, their retention time and Quantitation ions

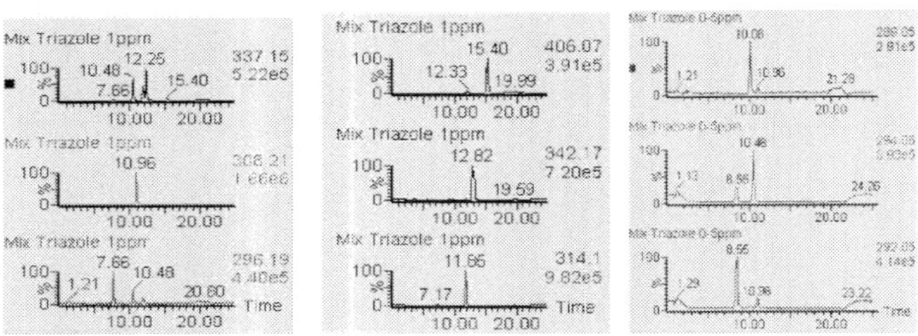


**Figure 4.** Mass spectra of triazoles; Fenbuconazole (337.1), Difenoconazole (406.1), Myclobutanil (289.1), Tebuconazole (308.1), Propiconazole (342.1), Triadimefon (294.1), Triadimenol (296.1), Hexaconazole (314.1) and Cyproconazole (292.1 ).

### *3.2.1. Mass spectrometer tuning*

Before the chromatographic method was established, the mass spectrometer was tuned to optimize the conditions of parameter for both the formation and detection of ions during an analysis. It is also done to increase the sensitivity and to optimize the mass peak resolution for the application. Optimization of both the ionization process and ion transportation in the mass spectrometer is important to achieve high sensitivity and selectivity and low detection limits in liquid chromatography / atmospheric pressure chemical ionization spectrometry (LC/APCI-MS) analysis. The optimization was done by changing one-variable-at-a time while the others are kept constant.

The mass spectrometer tuning was done using two methods; by infusing a sample with the syringe pump and also from the syringe pump into the LC flow line. This is to see the effect of mobile phase flow rate and composition on signal intensity and to allow optimization of the source parameters without making numerous injections in order to achieve parameters giving the highest sensitivity. Infusion experiments were carried out to examine the ionization and fragmentation patterns of the analytes. The instrument parameter; corona voltage, cone voltage, desolvation flow and temperature, cone flow and mass resolution were optimized to provide the best possible sensitivity by infusion. Corona voltage, was studied in the range from 3.5 V to 5 V and cone voltage studied in the range from 25 V to 35 V.

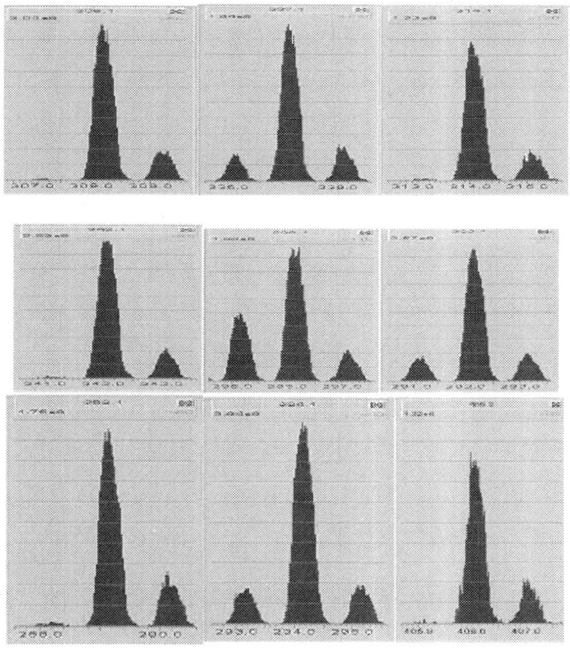

**Figure 5.** Mass of triazoles after tuning; (308.1), Fenbuconazole (337.1), Hexaconazole (314.1), Propiconazole (342.1), Triadimenol (296.1), Cyproconazole (292.1), Myclobutanil (289.1), Triadimefon (294.1) and Difenoconazole (406.2).

## 3.3. Sample preparation and extraction

### *3.3.1. Sample preparation*

The analysis is performed on a subsample of the laboratory sample, after appropriate comminution. This is to ensure the sub-sample is representative of the original laboratory samples.

In this study, 1-2 kg laboratory samples; cabbage, carrot, tomato and orange were used as representative samples. The laboratory samples were processed in a large chopper (Robot Coupe R5 V.V) and were blended to a consistent texture. Then 200g of the comminuted samples was transferred to another container and homogenized with the OMNI mixer homogenizer until homogeneous. This step is taken so that the 30 g samples taken for extraction are highly representative of the initial sample. Well comminuted samples can improve the shaking based extraction and less time is spent on the overall homogenization of the large initial laboratory samples [20]. An extremely homogeneous sample also maximizes surface area and ensures better extraction efficiencies.

30 g sub-sample was used by Aysal et al., 2007 [7] based on a study by Maestroni et al. 2000 [21-22] that showed results produced using the same chopper in the same laboratory gave representative results within generally ≤ 8% relative error of the mean concentration of the original sample.

### 3.3.2. *Sample extraction*

In contrast with acetone and acetonitrile-based methods, in which SPE is commonly employed, it has been reported only occasionally or no clean-up for ethyl acetate-based methods; however in this study dispersive-SPE clean-up was performed. Mol, H.G.J et al., 2007 [23] showed that laborious steps in multi residue analysis can be replaced by more efficient alternatives including the clean-up process. Solid-phase extraction previously used in the clean-up procedure which involves less dilution and is less laborious can be replaced by dispersive SPE, as described by Anastassiades et al., 20 [6].

SPE clean up used plastic cartridges containing various amounts of sorbent material and the procedures involve conditioning, sample transfer, elution, and evaporative re-concentration [23]. For this study, in the dispersive-SPE clean-up, 0.25 g primary secondary amine (PSA) and 1.5 g of anhydrous magnesium sulfate ($MgSO_4$) were added to a 10 mL aliquot of the sample extract and the mixture is mixed using a vortex mixer to evenly distribute the SPE material and facilitate the clean up process. The sorbent is then separated by centrifugation and the supernatant is ready for analysis. The function of the sorbent is to retain matrix components and not the analytes of interest. In some instances, other sorbents or mixed sorbents can be used depending on the samples and analytes.

A difficulty that was encountered by using ethyl acetate is that some of the most polar pesticides do not readily partition into ethyl acetate. It co-extracts with lipids and waxes, giving lower recoveries for the acid-base pesticides, it is sufficiently polar to penetrate into the cells of the matrix and it dissolves a great number of polar pesticides and their metabolites. On the other hand, ethyl acetate is partially miscible with water and the advantage is that it makes the addition of other non-polar solvents to separate water from the extract unnecessary. To increase the recoveries of polar compounds, large amounts of sodium sulfate ($Na_2SO_4$) are usually added in the procedures using ethyl acetate to bind the water. Polar co-solvents, such as methanol and ethanol, have been used to increase the polarity of the organic phase [23-26].

Different types of samples have different pH values that can affect the recoveries of pH-susceptible pesticides and their stability in the extracts. Therefore the pH of the extracts for some samples must be controlled [6, 20, 27]. Most pesticides are more stable at lower pH. Problematic pesticides that are strongly protonated at low pH the extracts must be buffered in the range of pH 2-7 [28]. The pH at which the extraction is performed can also influence the co-extraction of matrix compounds and pesticide stability. The pH of the samples extracted in this study was between pH 2.5 – pH 4.0. Sodium hydrogen carbonate ($NaHCO_3$) was added in the method to give a consistent pH during extraction independent of the initial sample pH.

Aysal P. et al., 2007 [7] mixed the sample 1:1 (w:w) with anhydrous sodium sulfate ($Na_2SO_4$) and used a 2:1 (v:w) ethyl acetate : sample ratio because it had been evaluated previously to achieve high recoveries. It resulted in good extraction efficiency and is practical with regard to achieving phase separation and avoidance of emulsions. [23-25].

The two conditions most relevant to extraction efficiency are the sample-to-solvent ratio and the addition of salt, which in ethyl acetate-based multi-residue methods has always been sodium sulfate. A study done by Mol, H.G.J et al. in 2003 [23] showed that the addition of salt improves the extraction efficiency for polar pesticides.

### *3.3.3. Dispersive-SPE clean-up*

The purpose of salt addition is to induce phase separation. The salting-out effect also influences analyte partition, which is dependent upon the solvent used for extraction. The concentration of salt can influence the percentage of water in the organic phase and can adjust its "polarity". In the QuEChERS method, acetonitrile alone is often sufficient to perform excellent extraction efficiency without the need to add non-polar co-solvents that dilute the extract and make the extracts too non-polar. By using deuterated solvents in the nuclear magnetic resonance studies, Anastassiades and colleagues [6] investigated the effect of various salt additions on the recovery and other extraction parameters. They studied the effect of polarity differences between the two immiscible layers. The use of magnesium sulfate as a drying salt to reduce the water phase helped to improve recoveries by promoting partitioning of the pesticides into the organic layer. To bind a significant fraction of water, the amount of magnesium sulfate exceeded the saturation concentration. The supplemental use of sodium chloride helps to control the polarity of the extraction solvents and thus influences the degree of matrix clean up of the QuEChERS method but too much of this salt will reduce the organic layer's ability to partition polar pesticides.

Dispersive solid-phase extraction is similar in some respects to matrix solid-phase dispersion developed by Barker [28-29] but in this instance, the sorbent is added to an aliquot of the extract rather than to the original solid sample as in matrix solid-phase extraction. In dispersive solid-phase extraction, a smaller amount of sorbent is used only because an aliquot of the sample is subjected to the clean up. Compared with SPE, dispersive solid-phase extraction takes less time and uses less labour and lower amounts of solvent without the extra steps such as channeling, analyte or matrix breakthrough, or preconditioning of SPE cartridges. Just as a drying agent is sometimes added to the top of an SPE cartridge, magnesium sulfate is added simultaneously with the SPE sorbent to remove much of the excess water and to improve the analyte partitioning to provide better clean up.

### 3.4. Quantitative determinations

All samples were quantified using the method of external standards. The linear concentration range was derived from the calibration graphs. Seven-point calibration curves for each compound were found to be linear ranging from 0.05 to 2.0 μg/ml, with 1/x weighting and with correlation coefficients ($r^2$) of >0.995. SIM traces were integrated for quantitation purposes. The limit of detection (LOD) and limit of quantification (LOQ) were estimated from the computer-generated software using a signal-to noise ratio (S/N) program. The LOD and LOQ were determined based on a signal to noise ratio of 3 and the limit of quantification (LOQ) was based on a signal to noise ratio of 10.

The internal standard was evaluated qualitatively only to confirm the injection of the sample extract. Normalization against the internal standard was not considered feasible because of unpredictable and varying matrix effects for several of the matrices studied in this work. A matrix-matched standard was also prepared by spiking the final extract; a fortification standard was added to the blank sample that had been extracted using the same procedure.

### 3.5. Pesticide recoveries

Recovery of pesticides from the fortified samples was calculated relative to that from a solvent standard and a matrix-matched standard. The acceptable percentage ranges for recovery (accuracy) and CV (precision) was based on CODEX criterion for method validation.

Recoveries and coefficients of variation of triazoles from fortified orange samples are shown in Table 2 and Fig. 6. Samples were spiked at 0.05mg/kg, 0.5mg/kg and 1mg/kg. The recoveries for these triazoles were from 60% to 145% with CV of 2.3 to 13.1% Most of the compound recoveries give more than 70% and fufills the codex acceptable recovery range. The recovery for Cyproconazole, tebuconazole and propiconazole at 1.0 mg/kg and myclobutanil at 0.05 mg/kg falls outside the acceptable range but the CV is within the acceptable range. Overall average recovery was 95% at all 3 fortification levels and all compounds met the CODEX CV acceptable range.

| | **Orange** | | |
|---|---|---|---|
| | **1.0 mg kg$^{1}$** | **0.5 mg kg$^{-1}$** | **0.05 mg kg$^{-1}$** |
| Triadimenol | 9.3 (2.3) | 59.6 (7) | 89.1 (11.4) |
| Cyproconazole | 128.8 (4.9) | 88.9 (4.7) | 75.2 (4.9) |
| Myclobutanil | 103.0 (3.9) | 83.9 (5.5) | 144.9 (4.8) |
| Triadimefon | 89.7 (2.7) | 65.7 (7.1) | 85.9 (3.1) |
| Tebuconazole | 142.4 (5.5) | 98.5 (4.0) | 100.9 (5.3) |
| Fenbuconazole | 111.4 (3.9) | 73.1 (5.9) | 91.84 (13.1) |
| Hexaconazole | 117.0 (7) | 87.0 (5.4) | 74.8 (6.4) |
| Propiconazole | 121.5 (3.2) | 75.0 (4.7) | 69.4 (9.8) |
| Difenoconazole | 118.7 (4.3) | 86.2 (5.9) | 87.7 (9) |

**Table 2.** Recovery of Triazoles in orange (n=5)

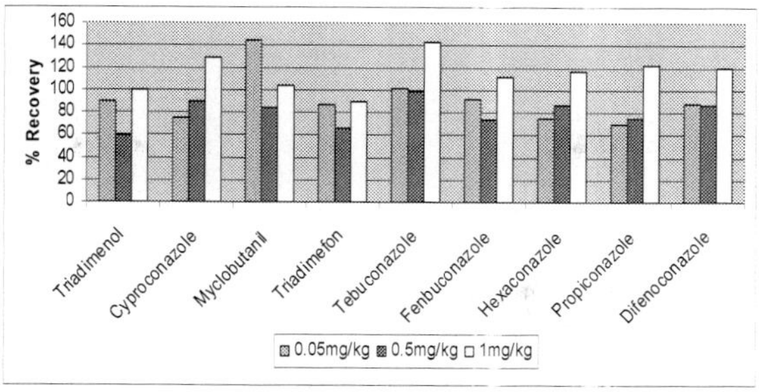


**Figure 6.** Recoveries fortified in Orange (n=5 at each level).

The recoveries and coefficients of variation of triazoles from fortified cabbage samples are shown in Table 3 and Fig. 7. Samples were spiked at 0.05mg/kg, 0.5mg/kg and 1mg/kg. The recoveries for 9 triazoles were within the acceptable range for 1 mg/kg fortification levels. They ranged between 98-120% (CV of 2.6% to 7.1%). Recoveries obtained at 0.5 mg/kg were in the range of 76-100% with CV of 4.1 to 18.9%). The recoveries for cyproconazole and fenbuconazole at 0.5 mg/kg were in the acceptable range but CV% was out of range. Recoveries for cabbage spiked at 0.05mg/kg were between 53-98%. A lower recovery was obtained for cyproconazole and tebuconazole while fenbuconazole was almost all lost (0.7%) and the CV was also so high (149.9%). The overall average recovery was 91%.

| | Cabbage | | |
|---|---|---|---|
| | **1.0 mg kg$^{1}$** | **0.5 mg kg$^{-1}$** | **0.05 mg kg$^{1}$** |
| Triadimenol | 115.8 (2.6) | 99.0 (4.1) | 88.7 (2.5) |
| Cyproconazole | 98.6 (7.1) | 76.0 (16.8) | 53.5 (13.0) |
| Myclobutanil | 119.9 (2.7) | 98.8 (6.3) | 92.1 (4.1) |
| Triadimefon | 112.9 (6.0) | 100.3 (5.1) | 101.9 (2.3) |
| Tebuconazole | 98.9 (3.4) | 77.8 (10.0) | 58.9 (6.9) |
| Fenbuconazole | 101.8 (6.1) | 76.5 (18.9) | 0.7 (149.4) |
| Hexaconazole | 115.6 (2.7) | 94.2 (8.3) | 71.5 (3.8) |
| Propiconazole | 113.15 (4.3) | 96.6 (4.8) | 96.4 (3.6) |
| Difenoconazole | 112.4 (5.5) | 95.6 (8.1) | 98.3 (9.9) |

**Table 3.** Recovery of Triazoles in cabbage (n=5)

The recovery results obtained from spiked tomato at different levels was between 66-144% (see Table 4). The CV for all compounds at different fortification levels were in the range of 2.6 – 11.9% and met the CV acceptable range except for fenbuconazole (16.9%). The recovery for propiconazole at all the 3 concentrations of 0.05 mg/kg and 0.5 mg/kg spiked did not meet the acceptance limit but the CV's met the acceptable limit. The overall average recovery was 105.3%.

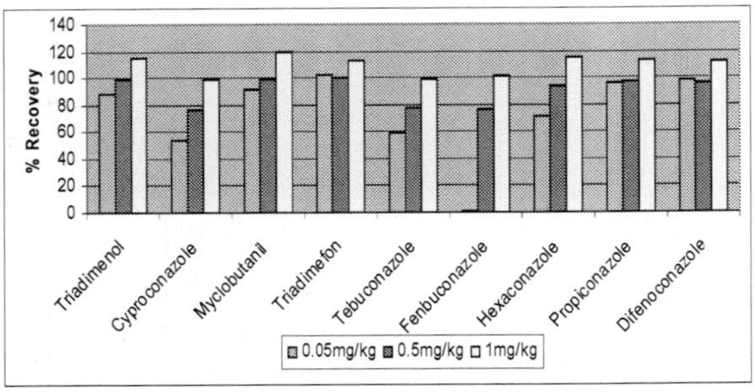


**Figure 7.** Recoveries fortified in Cabbage (n=5 at each level)

| | Tomato | | |
|---|---|---|---|
| | **1.0 mg kg$^{1}$** | **0.5 mg kg$^{-1}$** | **0.05 mg kg$^{-1}$** |
| Triadimenol | 105.7(10.3) | 95.7 (8.9) | 87.8 (9.1) |
| Cyproconazole | 97.1 (4.2) | 88.8 (8.3) | 92.6 (11.9) |
| Myclobutanil | 91.9 (8.3) | 88.0 (8.6) | 66.2 (10.9) |
| Triadimefon | 103.6 (9.1) | 96.4 (6.3) | 118.9 (10.7) |
| Tebuconazole | 105.6 (8.8) | 88.9 (8.9) | 115.2 (10.3) |
| Fenbuconazole | 104.0 (2.6) | 93.8 (5.8) | 103.4 (16.9) |
| Hexaconazole | 108.3 (9.4) | 86.6 (10.0) | 84.9 (8.0) |
| Propiconazole | 188.0(10.2) | 144.7 (4.6) | 175.6 (7.1) |
| Difenoconazole | 107.1(15.7) | 99.7 (9.7) | 105.2 (10.7) |

**Table 4.** Recovery of Triazoles in tomato (n=5)

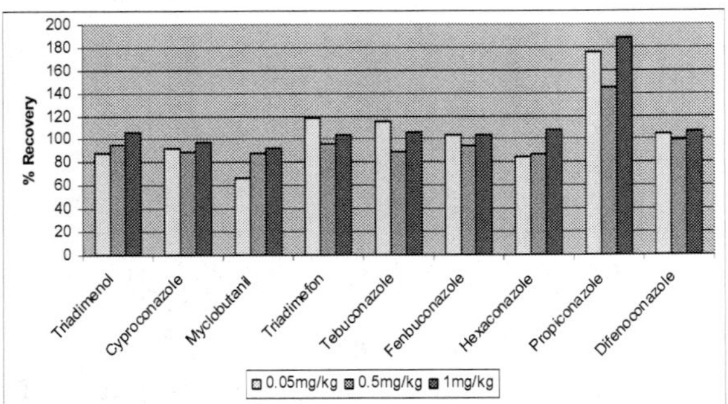


**Figure 8.** Recoveries fortified in tomato (n=5 at each level)

The recoveries for all spiked triazoles at 3 concentration levels in carrot are shown in Table 5 and Figure 9. As shown in Table 5, the recoveries for carrot spiked at 0.05 mg/kg were quite low for most of the analytes, between 48-111% with CV in the range of 5.4 to 56.6%. The recovery spiked at 0.5 mg/kg was between 68-85% but difenoconazole recovery was very high at 151%. Contrary to the recovery spiked at 0.05 mg/kg, recoveries spiked at 1 mg/kg were very high (more than 100%) for most compounds between 86-142% and CV for all compounds did not meet the range. The overall average recovery was 90.7% for all fortification levels.

| | Carrot | | |
|---|---|---|---|
| | **1.0 mg kg$^{-1}$** | **0.5 mg kg$^{-1}$** | **1.0 mg kg$^{-1}$** |
| Triadimenol | 86.0 (15.2) | 69.8 (7.4) | 48.3 (6.1) |
| Cyproconazole | 113.0 (14.2) | 72.9(11.0) | 64.7 (8.6) |
| Myclobutanil | 126.0 (14.4) | 75.3 (8.2) | 54.7 (5.4) |
| Triadimefon | 121.0 (18.8) | 68.6 (8.2) | 50.4 (10.8) |
| Tebuconazole | 142.0 (15.5) | 75.5 (7.3) | 73.9 (46.7) |
| Fenbuconazole | 109.0 (14.3) | 74.4 (8.0) | 51.0 (7.4) |
| Hexaconazole | 115.0 (16.2) | 85.5 (8.6) | 66.5 (6.6) |
| Propiconazole | 137.0 (14.9) | 72.3 (7.1) | 110.6 (56.6) |
| Difenoconazole | 121.0 (15.8) | 151.0 (6.8) | 111.4 (6.1) |

**Table 5.** Recovery of Triazoles in carrot (n=5)

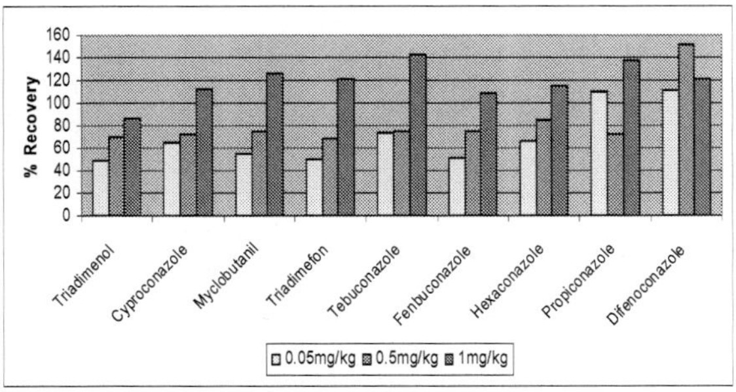


**Figure 9.** Recoveries fortified in Carrot (n=5 at each level)

## 4. Conclusion

Lower recoveries for some analytes in certain matrices and at certain concentrations in pesticide residue analysis could be due to the degradation of base sensitive pesticides in higher pH samples, or degradation of acid sensitive pesticides in lower pH samples. And protonization of basic pesticides in acidic conditions reduces partition into organic layer.

For consistent and higher recoveries, some considerations that we need to look into are the homogeneity of the samples, the choice of solvent, sorbent(s) and salt(s) used during clean-up process. The EtOAc modified QuEChERS method was demonstrated to provide consistent and reproducible recoveries for tomato, cabbage and orange for most triazole compounds but not for carrot. High Performance Liquid Chromatography coupled with mass spectrometry by atmospheric pressure ionisation can be used for the identification of triazole fungicides in vegetables and additional confirmatory is not needed.

More sensitive analytical methods such as LC-QTof (Liquid chromatography high resolution time-of-flight), orbitrap mass analyzers (LC-HR-MS), LC-MS/MS that have higher sensitivity and specificity can also be used. LC-QTof, LC-MS/MS can be used for screening and confirmation work and need no additional confirmatory. Liquid chromatography coupled with high resolution time-of-flight or orbitrap mass analyzers (LC-HR-MS) seems to open new and attractive possibilities for residue analysis [30-31].

## Author details

Nor Haslinda Hanim Bt Khalil
*Food Quality Laboratory, Health And Environmental, Dept., Kuala Lumpur City Hall, Selayang, Kuala Lumpur*

Tan Guan Huat
*Dept. of Chemistry, Faculty of Science, Universiti Malaya, Lembah Pantai, Kuala Lumpur*

## Acknowledgement

The author wish to thank the Director and Deputy Director of Health and Environmental Department, Kuala Lumpur City Hall for their support of this project and Mohd. Fairuz, Assistant Science Offcer of Food Quality Laboratory, Kuala Lumpur City Hall for the technical assistance.

## 5. References

[1] Wu Y. S.; Lee H. K.; and Li S. F. Y. (2001) *J Chromatography. A.*, 912, 171-179

[2] Mueller, 2006]

[3] Fillion, J., Sauvé, F., and Selwyn, J. (2000). Multiresidue method for the determination of residues of 251 pesticides in fruits and vegetables by gas chromatography/mass spectrometry and liquid chromatography with fluorescence detection. *J. AOAC Int.* 83, 698-713.

[4] S. Nemoto; K. Sasaki; S. Eto; I. Saito; H. Sakai; T. Takahashi; Y. Tonogai; T. Nagayama; S. Hor; Y. Maekawa; and M. Toyoda (2000). *J. Food Hyg. Soc. Japan* 41, 233–241 (in Japanese).

[5] H. Obana; K. Akutsu; M. Okihashi; and S. Hori (2001). *Analyst* 126, 1529–1534

[6] Anastassiades, M., Lehotay, S.J., Štajnbaher, D., and Schenck, F.J. (2003). Fast and Easy Multiresidue Method Employing Acetonitrile Extraction /Partitioning and"Dispersive Solid-Phase Extraction" for the Determination of Pesticide Residues in Produce. *J. AOAC Int.* 86 (2), 412-431.

[7] Aysal, P., Ambrus, Á., Lehotay, S.J.; and Cannavan, A. (2007). Validation of an efficient method for the determination of pesticide residues in fruits and vegetables using ethyl acetate for extraction. *Journal of Environmental Science and Health B,* in press (Vol.B42, No.5)

[8] Food and Drug Administration (1999) *Pesticide Analytical Manual Volume I*: Multiresidue Methods, 3rd Edition, U.S. Department of Health and Human Services, Washington, DC.

[9] Luke, M. A., Froberg, J.E., and Masumoto, H. T. (1975). Extraction and cleanup of organochlorine, organophosphate, organonitrogen, and hydrocarbon pesticides in produce for determination by gas-liquid chromatography. *J. Assoc. Off. Anal. Chem.* 58, 1020-1026.

[10] Cook, J., Beckett, M. P., Reliford, B., Hammock, W., and Engel, M. (1999) Mutiresidue analysis of pesticides in fresh fruits and vegetables using procedures developed by the Florida Department of Agriculture and Consumer Services. *J.AOAC Int.* 82, 1419-1435.

[11] Lee, S. M., Papathakis, M. L., Hsiao-Ming, C. F., and Carr, J. E. (1991). Multipesticide residue method for fruits and vegetables: California Department of Food and Agriculture. *Fresenius J. Anal. Chem.* 339, 376-383.

[12] Andersson A., and Pålsheden H. (1991). Comparison of the efficiency of different GLC multi-residue methods on crops containing pesticide residues. *Fresenius J.Anal. Chem.* 339, 365-367.

[13] Sheridan, R. S., and Meola J. R. (1999). Analysis of pesticide residues in fruits, vegetables, and milk by gas chromatography/tandem mass spectrometry. *J. AOAC Int.* 82, 982-990.

[14] Lehotay, S. J. (2000). Determination of pesticide residues in nonfatty foods by supercritical fluid extraction and gas chromatography/mass spectrometry: collaborative study. *J. AOAC Int.* 83, 680-697.

[15] Wylie, P; and Chin, K.M (2001). *Labplus International* – September/October

[16] Aguilar, C; Ferrer, I.; Borrul, F.; Marce, R.M. (1998). *J. Chromatography. A,* 794 (1/2), 147-164

[17] Aguilar, C; Penalver, S.; Pocurull, E. Borrull, F.; Marce, R.M. (1998). *J. Chromatography. A,* 795 (1), 105-116

[18] Lehotay,S.J.; Kyung, A.S.; Kwon, H.; Koeksukwiwat,U.; Fu, W.; Mastrovska, K.; Hoh, E.; and Leepipatpiboon, N. (2010). *Journal of Chromatography A.* 1217, 2548-2560.

[19] Guzetta, A (2001). Reverse Phase HPLC Basics for LC/MS, An Ion Source tutorial.

[20] Lehotay, S.J.; de Kok, A.; Hiemstra, M.; and van Bodegraven, P. ( 2005). Journal of AOAC International. Vol. 88, No.2,

[21] Maestroni, B.; Ghods, A.; El-Bidaoui, M.; Rathor, N.; Ton, T.; Ambrus, *A. (*2000). In *Principles and Practices of Method Validation;* Fajgelj, A., Ambrus, A.,Eds.; The Royal Society of Chemistry, Cambridge, England, 49-58.

[22] Maestroni, B.; Ghods, A.; El-Bidaoui, M.; Rathor, N.; Jarju, O.P.;Ton, T.; Ambrus, A. (2000). In *Prin-ciples and Practices of Method Validation;* Fajgelj, A., Ambrus, A.,Eds.; The Royal Society of Chemistry: Cambridge, England. 59-74

[23] Mol HGJ et al 2007

[24] Kadenczki, L.; Zoltan, A.: Gardi, I.; Ambrus, *A.*: Gyorfi, L.; Reese, G.; Ebing, W. (1992). *J. AOAC Int.*, 75, *(1)*, 53-61.

[25] Holstege D.M. et al., 1994

[26] Holland, P.T.; Boyd, A.J.; Malcolm, C.P. (2000) In *Principles and Practices of Method Validation;* Fajgelj, A., Ambrus, A., Eds.; The Royal Society of Chemistry: Cambridge, England,; 29-40.

[27] Lehotay, S.J. (2007). *J. AOAC mt. 90,* 485-520.

[28] S.A. Barker, LCGC, Special Supplement, Review on Modern Solid-Phase Extraction (May 1998)

[29] S.A. Barker (2000). J. Chromatogr. A. 880, 63-68

[30] Thurman E.M; Ferrer, I,; and Fernández-Alba A.R. (2005). J Chromatogr A. 1067, *127-34*

[31] Picó, Y; and Barceló, D (2008). *Trends Anal. Chem*. 27, 821-835

Chapter 9

# Effect of Simulated Rainfall on the Control of Colorado Potato Beetle (Coleoptera: Chrysomelidae) and Potato Leafhopper (Homoptera: Cicadellidae) with At-Plant Applications of Imidacloprid, Thiamethoxam or Dinotefuran on Potatoes in Laboratory and Field Trials

Gerald M. Ghidiu, Erin M. Hitchner and Melvin R. Henninger

Additional information is available at the end of the chapter

http://dx.doi.org/10.5772/55331

## 1. Introduction

The Colorado potato beetle, *Leptinotarsa decemlineata* (Say) [Order Coleoptera] (CPB), is considered the most important pest of potatoes throughout the northeastern and mid-Atlantic regions of the United States [1]. Both larval and adult CPB feed on potato foliage, stems, and flowers, which can severely defoliate plants, significantly reducing yields [2]. Growers rely on insectidicides to control CPB in the field but it has developed resistance to 52 different compounds used against it, which includes all major insecticide classes [3].

Another economically important insect pest of potatoes is the potato leafhopper, *Empoasca fabae* (Harris) [Order Homoptera] (PLH), a sap-feeding insect that causes damage known as "hopper burn" [4]. Feeding results in curling, stunting, yellowing and eventual browning of the potato foliage, and even low numbers of leafhoppers can cause significant yield losses [4, 5]. The potato leafhopper overwinters in the southern US and migrates northward on wind currents [6,7], typically arriving in the northeast United States in mid- to late May or early June each year.

Currently throughout the potato-producing regions of the United States, the neonicotinoid class of insecticides, which includes dinotefuran, imidacloprid and thiamethoxam, is widely used for control of several insect pests of white potatoes [8,9], including early to mid-season infestations of both CPB and PLH. These insecticides are toxic to both of these pests [10], have excellent systemic uptake and translocation in plants, are used at low application rates, and present reduced environmental hazards [11,12]. Although most neonicotinoids can be applied as either a seed treatment, a soil treatment or as a post-planting foliar treatment, many commercial growers prefer a soil application at planting time. Applied to the soil, these materials are absorbed by the roots and translocated acropetally (xylem movement) within the whole plant [12,13]. However, the effectiveness of soil-applied insecticides are influenced by soil and climatic conditions, including soil moisture, clay and organic content, rainfall, soil temperature, plant size, sorption, adsorption, and physical properties of the insecticide, such as stability to chemical and microbial degradation [14,15,16]. Thus, when used as a soil-applied insecticide, the length of effective protection can vary and additional foliar insecticide sprays may be needed to control midsummer populations of both the CPB and PLH.

The solubility of an insecticide and its activity in soil and uptake by the roots is known to be important, but this relationship has not been well documented [17]. In general, increased solubility is positively related to increased uptake of a systemic chemical. A compound's relative affinity between soil and water phases is measured by log Koc (organic carbon referenced sorption coefficients). With sufficient soil moisture, materials with higher log Koc bind more tightly to soil colloids and release at a slower rate, affecting the availability for root uptake; materials with a higher solubility enter the plant roots more quickly and thus move through a plant faster than materials with a lower solubility. The uptake of dinotefuran, which is 80 times more soluble than imidacloprid, in both yellow sage and poinsettia plants was more rapid and resulted in quicker and higher percentage mortality of whitefly nymphs compared with imidacloprid [18]. However, materials with a lower log Koc are also more easily leached out of the root zone, which can reduce effectiveness. For example, thiamethoxam has a strong potential to leach under heavy rainfall conditions [19], and imidacloprid has a potential to leach to ground water [20]. Leaching of the insecticide would impact insect pest control because less insecticide would be available for uptake by the plant, resulting in reduced efficacy or in reduced longevity of efficacy. Although much research has been conducted on the bioefficacy of the neonicotinoids against CPB and PLH as soil-applied insecticides, little published information is available on the effect of rainfall on these materials after application to the soil.

Therefore, laboratory and field studies were conducted to examine the effect of two levels of simulated rainfall on three neonicotinoids with different solubilities for control of CPB and PLH on white potato.

## 2. Methods and materials

**Laboratory Trials.** Two "Superior" cv. potato seed pieces were each planted into 22.9 cm (9") diam pots half-filled with soil in the field on 23 Apr. The pots were then aligned in the field in a straight row, each pot touching the next, to approximate a grower-practiced seed spacing of 22.9 cm (9"). Treatments consisted of three neonicotinoid insecticides, at full labeled rates, and an untreated: imidacloprid (Admire PRO, 635.9 ml/ha [8.7 fl oz/acre], BayerCropScience, Research Triangle Park, NC), thiamethoxam (Platinum 2SC, 584.2 ml/ha [8.0 fl oz/acre], Syngenta, Greensboro, NC), dinotefuran (Venom 70SG, 525.5 g/ha [7.5 oz/acre], Valent BioSciences, Libertyville, IL), and an untreated water spray. The solubility of the three neonicotinoids, from least to most soluble, are imidacloprid (solubility of 0.51 g/liter at 20$^0$ C, [20], thiamethoxam (solubility of 4.1 g/liter at 20$^0$ C, [12], and dinotefuran (solubility of 40.0 g/liter at 20$^0$ C, [21]. To simulate a field application, treatments of insecticides were applied over the top of the pot as a 10.2 cm (4") band open-furrow application on 23 Apr using a 3785 ml (1 gal) Agway (Southern States Cooperative, Richmond, VA) water sprinkler calibrated to deliver 189.3 liter/0.405 ha (50 gpa). The seed pieces were then covered by filling the pots with clean field soil. Because soil texture, clay content and organic matter influence mobility, the same soil (Sassafrass sandy loam, 65% sand, 23% silt, 12% clay, 1% organic matter) was used for all the laboratory pots and for the field trials. A total of 32 pots were prepared (8 pots per treatment with the 4 treatments described above) and were immediately moved to an environmentally-controlled plexiglass greenhouse. Pots were evenly divided into 2 groups, with 4 pots (4 replications) of each treatment in the low rainfall group and 4 pots of each treatment in the high rainfall group. Thus main plots were amount of rainfall, and sub-plots were insecticide treatment. All 16 pots in either the low rainfall group or high rainfall group were placed on an 1.8 m (6 ft) diameter round table. A mechanical rainfall simulator [see 22] consisting of a rotating boom with a single TeeJet 8010 nozzle, operated by a 0.254 metric hp (¼ hp) electrical motor, delivered a 20.3 cm (8") band of water over the pots to simulate rainfall. Two rain gauges were placed opposite each other on the table between the pots: once every Monday the low rainfall group received 1.3 cm water (0.5") as measured in the rainfall gauge, and every Monday and Thursday the high rainfall group received 3.8 cm (1.5") each time, for a total of 7.62 cm (3") per week. Simulated rainfall commenced on 26 Apr, 3 days after planting, and continued every week until 11 Jun (a period of 8 weeks for a total of 4" low rainfall regimen, and 24" high rainfall regimen). Two leaves from the middle 1/3 foliage from each plant in each pot were picked and placed in sterile 12.7 cm diameter (5") Petri dishes on 30 May, 3, 15, 25 Jun and 2 Jul. Just after leaves were placed in the Petri dishes, CPB larvae were collected from untreated nursery potato plots ("Superior" cv) in the field and five same-instar larvae were placed in each Petri dish with the leaves and placed in the laboratory (2$^{nd}$ and 3$^{rd}$ instar larvae were tested May through early Jun, 3$^{rd}$ and 4$^{th}$ instar larvae were tested mid- to late Jun, and adults were tested early Jul because no larvae were available). After 72 hr, the total percentage leaf-feeding, based on visual assessment of percentage leaf tissue consumed by the CPB, and the number of live CPB in each Petri dish was recorded (3, 8, 21, 28 Jun and 5 Jul).

**Field Trials.** "Superior" cv. white potatoes were planted on 10 Apr into a prepared (disked, limed and fertilized) Sassafras sandy loam field. This soil was the same as used in the laboratory trials previously described. Plots consisted of 3 rows of potatoes, each row 7.62 m long (25 ft) and 0.9 m wide (3 ft), replicated 4 times in a split-plot design: whole plots were amount of rainfall (a rainfall regimen of either 1.3 cm/week [½″] or 7.62 cm/week [3″), and sub-plots were insecticide treatment (imidacloprid, thiamethoxam, dinotefuran, or untreated). Insecticide treatments were the same as in the laboratory trials previously described and were applied at the same rates to the furrow at planting using a hand-held 7.6 liter (2-gal) Agway sprinkler can calibrated to deliver 189.3 liter/0.405 ha (50 gpa) in a 10.2 cm (4-in) band, after which the furrows were immediately closed; one treatment consisted of no insecticide. Whole plots were irrigated with an overhead irrigation system consisting of a 5.1 cm (2″) main water pipe connected to a Rainbird J-20 revolving irrigation head delivering 0.5 cm water per 0.405 ha (0.2″ per acre) per hour. Each whole plot had two rain gauges at plant height, one located near the irrigation head and one at the furthest wet point away from the irrigation head. Plots received overhead irrigation every Wednesday starting 23 Apr and continued each week through 27 Jun (either 1.3 cm [1/2″]or 7.62 cm [3″]) over a total of 10 weeks. Plots that received a natural rainfall less then the treatment amount during the week received additional irrigation only to bring the total to 1.3 cm or 7.62 cm, and plots that received more natural rainfall than the treatment amount received no irrigation.

The total number of Colorado potato beetle larvae (small and large larvae) per 3 hills, and percentage plant defoliation caused by CPB feeding were recorded on 1, 11, 17, and defoliation ratings only on 25 Jun. Potato leafhopper damage ratings (0 = no damage, 5 = severe damage) were recorded on 25 Jun and 3 Jul.

All data were recorded from the center row of each 3-row plot. Data were averaged to obtain a plot mean for all recorded observation. Data for both the laboratory and field trial were subjected to a split-plot analyses of variance (ANOVA) [23]. Means were separated using Tukey's HSD Studentized range test [23] and were plotted on graphs with Microsoft Excel (www.microsoft.com/en-US/excel365/). .

## 3. Results

### 3.1. Laboratory results

The results of the ANOVA to test the main effects (amount of simulated rainfall and insecticide) and rainfall by insecticide interaction on CPB in laboratory trials are summarized in Table 1. ANOVA demonstrated significant ($P<0.05$) insecticide effects for both CPB mortality and percentage leaf feeding on all dates recorded. However, the main effect of the amount of simulated rainfall was significant only on the first date for percentage leaf tissue eaten by CPB larvae, and only on the first two dates for CPB larval mortality (Table 1). Similarly, the rainfall by insecticide interaction for CPB mortality was significant ($P<0.01$) only on the first observation date (Table 1, and significant ($P<0.01$) only on the first two observation dates for percentage leaf tissue eaten..

When CPB larvae were placed on leaves treated with imidacloprid or thiamethoxam, mortality was significantly ($P<0.01$) higher compared with larvae placed on leaves treated with dinotefuran or the untreated for the low rainfall regimen until 17 Jun (Fig. 1), when mortality of CPB on leaves treated with imidacloprid decreased; thiamethoxam remained

| | No. dead CPB/20 after 72 hr | | | | | % leaf tissue eaten by CPB after 72 hr | | | | |
|---|---|---|---|---|---|---|---|---|---|---|
| | 6/3 | 6/8 | 6/21 | 6/28 | 7/5 | 6/3 | 6/8 | 6/21 | 6/28 | 7/5 |
| MP | | | | | | | | | | |
| SP | * | * | ns | ns | ns | * | ns | ns | ns | ns |
| MPxS | ** | ** | ** | * | ** | ** | ** | ** | ** | ** |
| P | ** | ns | ns | ns | ns | ** | ** | ns | ns | ns |

MP= main effect of simulated rainfall; SP= main effect of insecticides; MPxSP= interaction
[1]Summary of results of analysis of variance. ns= nonsignificant; *, P<0.05%, **, P<0.01%

**Table 1.** Summary of effects of simulated rainfall and insecticides on efficacy of 3 soil-applied neonicotinoids against Colorado potato beetle (CPB) on white potatoes in laboratory trials[1], Bridgeton, NJ 2007

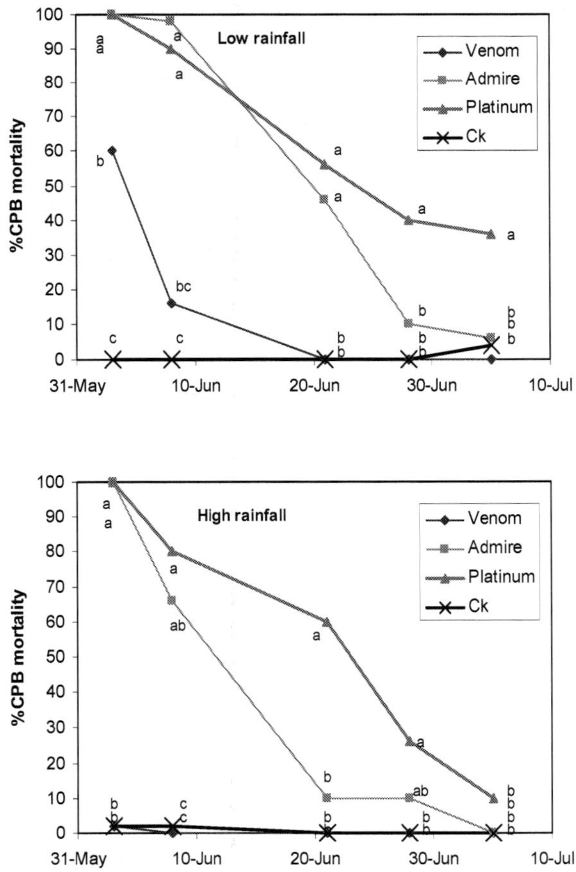


**Figure 1.** Effect of two levels of simulated rainfall on mortality of Colorado potato beetle larvae on potato foliage treated with neonicotinoids in a laboratory bioassay. Bridgeton, NJ. 2007.

significantly (P<0.01) more toxic to CPB larvae than all other treatments through 7 Jul. Under the high rainfall regimen, larvae placed on leaves treated with imidacloprid or thiamethoxam had a significantly (P<0.01) higher mortality on 28 May and 7 Jun than did larvae placed on leaves treated with dinotefuran or leaves from plants that received no insecticide; mortality of CPB on leaves treated with thiamethoxam was still significantly higher than that of CPB larvae on leaves treated with dinotefuran or the leaves that received no insecticide through the end of Jun (Fig. 1).

CPB larvae placed on leaves from potato plants treated with imidacloprid or thiamethoxam ate significantly (P<0.01) less leaf tissue than did larvae on potato leaves from plants treated with dinotefuran or on leaves with no insecticide on each day recorded for both the low and high rainfall regimen (Fig. 2). Leaves from potato plants treated with dinotefuran under the low rainfall regimen had significantly (P<0.01) less CPB feeding damage as compared with

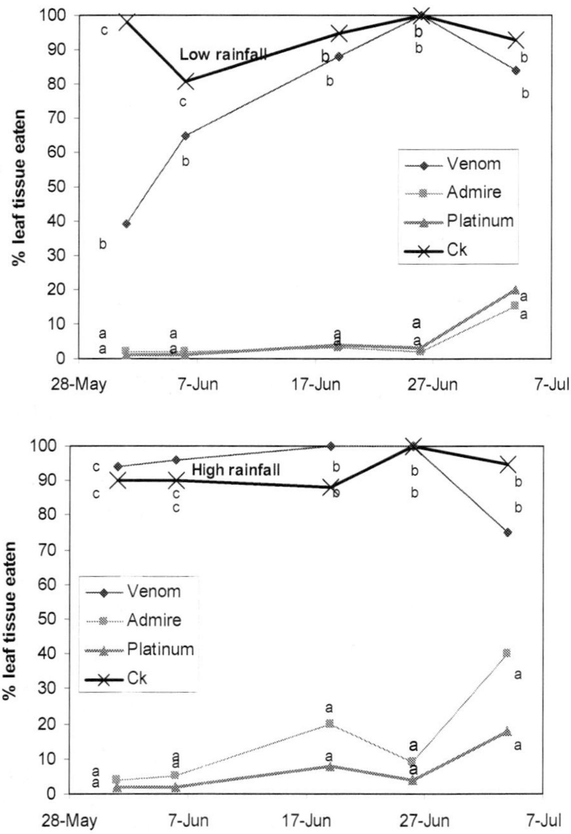


**Figure 2.** Effect of two levels of simulated rainfall on percentage leaf tissue eaten by Colorado potato beetle larvae on potato treated with neonicotinoids in a laboratory bioassay. Bridgeton, NJ 2007.

leaves with no insecticide only on 28 May and 7 Jun, but not on 17, 17 Jun or 7 Jul. Leaves from dinotefuran-treated potato plants under the high rainfall regimen showed no significant differences for damage as compared with the leaves with no insecticide on all dates of the bioassay (Fig. 2).

### 3.2. Field results

The results of the ANOVA to test the main effects (amount of simulated rainfall and insecticide) and rainfall by insecticide interaction on CPB and PLH in field trials are summarized in Table 2. ANOVA demonstrated significant (P<0.01) insecticide (sub-plots) effects for all data (total CPB, CPB defoliation and PLH damage rating). The main effect of the amount of simulated rainfall was significant (P<0.05) only for PLH damage. A significant (P<0.05) interaction of the amount of simulated rainfall by insecticide was observed for both the percentage defoliation caused by CPB and for PLH damage.

| | No. CPB/3 hills | | | % CPB defoliation | | | | PLH damage rating (0-5 scale) | |
|---|---|---|---|---|---|---|---|---|---|
| | 6/1 | 6/11 | 6/17 | 6/1 | 6/11 | 6/17 | 6/25 | 6/25 | 7/3 |
| MP | ns | ns | ns | ns | ns | ns | ns | ** | ** |
| SP | ** | ** | ** | ** | ** | ** | ** | ** | ** |
| MPxSP | ns | ns | ns | * | * | * | ns | * | * |

MP= main effect of simulated rainfall; SP= main effect of insecticides; MPxSP= interaction
[1]Summary of results of analysis of variance. ns= nonsignificant; *, P<0.05%, **, P<0.01%

**Table 2.** Summary of effects of simulated rainfall and insecticides on control of Colorado potato beetle (CPB) and potato leafhopper (PLH) on potatoes in field trials[1], Bridgeton, NJ 2007

Plots treated with either imidacloprid or thiamethoxam had fewer CPB larvae on all dates recorded as compared with plots that received no insecticide in both the high and low rainfall regimens; the number of CPB larvae/3 hills recorded in plots treated with imidacloprid or thiamethoxam remained low throughout the season (Fig. 3). The plots treated with dinotefuran had significantly (P<0.05) more CPB larvae then did either of the plots treated with either imidacloprid or thiamethoxam on 17 Jun. There were no significant (P<0.05) differences between plots treated with dinotefuran and plots that received no insecticide for number of CPB larvae/3 hills in either the low or high rainfall regimen on each day observed throughout the season. Both imidacloprid and thiamethoxam were effective in reducing the numbers of CPB larvae throughout the season (from planting through late Jun). Under the low rainfall regimen, dinotefuran was effective in reducing CPB larvae only through early Jun, and under the high rainfall regimen the number of CPB larvae/3 hills for potato plants treated with dinotefuran was not significantly (P<0.05) different than plants that received no insecticide on any date observed.

The percentage defoliation caused by CPB feeding was significantly (P<0.05) less in plots treated with imidacloprid or thiamethoxam as compared with dinotefuran or plots that received no insecticide after 12 Jun in both the low and high rainfall regimens (Fig. 4). The percentage defoliation in plots treated with dinotefuran was less than in plots that received no insecticide but greater than either imidacloprid or thiamethoxam in the low rainfall regimen, but was similar to the plots that received no insecticide in the high rainfall regimen, although these differences were not significant (P<0.05).

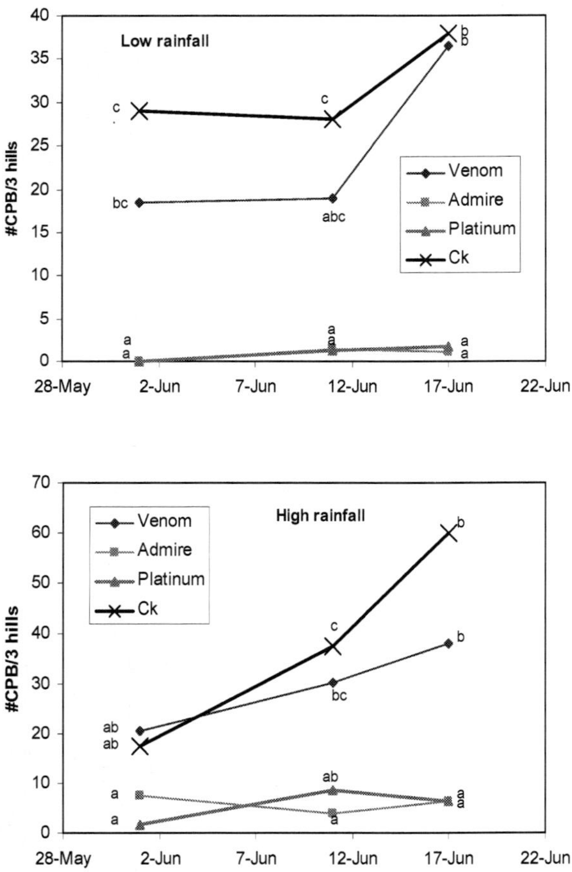


**Figure 3.** Effect of two levels of simulated rainfall on populations of Colorado potato beetle larvae on potatoes treated with neonicotinoid insecticides at planting, Bridgeton, NJ. 2007.

All insecticide-treated plots had significantly less (P<0.05) PLH damage on each day recorded for both the low and high rainfall regimens, as compared with plots that received no insecticide (Fig. 5). In the insecticide-treated plots that received low rainfall, no significant differences among the three treatments were recorded. Under the high rainfall regimen, however, the dinotefuran-treated plots had significantly higher PLH damage on 25 Jun than did plots treated with either imidacloprid or thiamethoxam, and by 7 Jul, plots treated with thiamethoxam had significantly (P<0.05) less PLH damage than all other treatments.

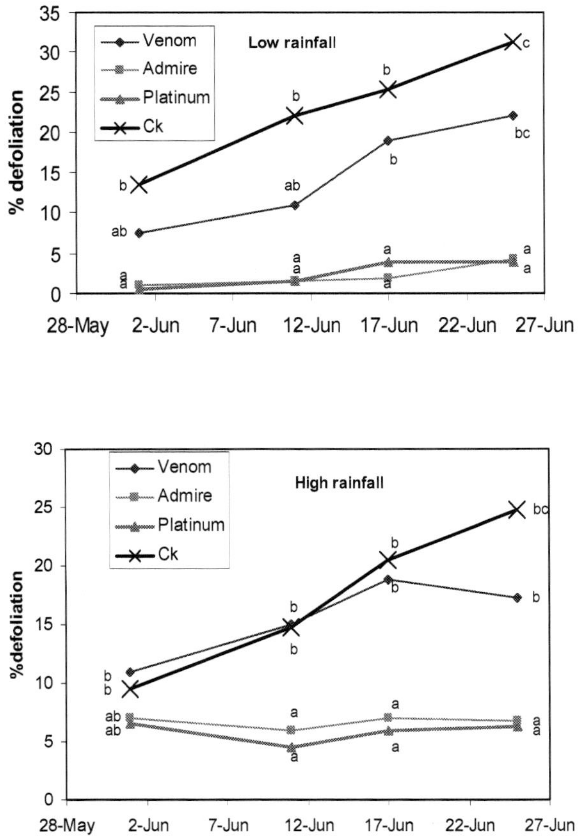


**Figure 4.** Effect of two levels of simulated rainfall on percentage plant defoliation caused by Colorado potato beetles on potatoes treated with neonicotinoids at planting, Bridgeton, NJ. 2007.

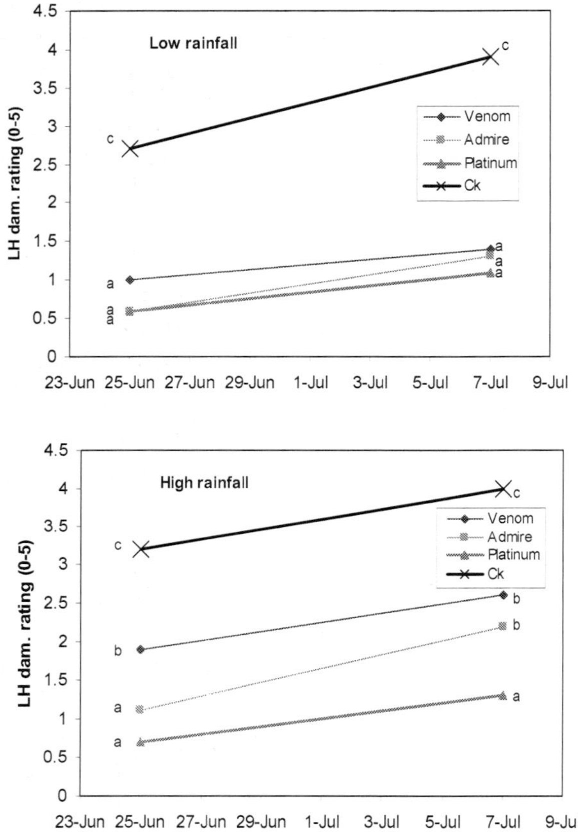


**Figure 5.** Effect of two levels of simulated rainfall on damage caused by potato leafhopper on potatoes treated with neonicotinoids at planting, Bridgeton, NJ. 2007.

## 4. Discussion

The CPB mortality bioassay showed that imidacloprid and thiamethoxam were more toxic to CPB larvae than dinotefuran during May and early June, but mortality of CPB larvae for all treatments declined over time (Fig. 1). It is likely that the continual simulated rainfall was partially responsible for this decrease in efficacy since dinotefuran, the most soluble material, was effective against CPB larvae early with the low amount of simulated rainfall, but was not effective at any time with the high amount of rainfall, suggesting that the material was leached from the soil before the potato plants had enough root matter to absorb the insecticide. Mortality would also decline over time because the material is no longer available in the foliage when the CPB were placed on the leaves in the petri dish, as the bioassay was conducted from 37 to 70 days after the in-furrow treatment applications. It is also possible that CPB mortality within all treatments declined over time because the laboratory test larvae were field-harvested from an untreated potato nursery, and only large larvae (3$^{rd}$ and 4$^{th}$ instars) were available in late June whereas only small larvae (1$^{st}$ and 2$^{nd}$ instars) were available in late May and early June, and only adults were available in July. Large CPB larvae and adults are generally more difficult to intoxicate than small larvae [24,25,26]. Further, mortality of CPB on leaves from plants treated with thiamethoxam was significantly (P=0.05) higher than that of all other treatments, including imidacloprid, under the low rainfall regimen on 28 Jun and 5 Jul, the last two test dates, suggesting that thiamethoxam remains more active in the plant over a longer period of time. Plants from soybean seeds that had been treated with with thiamethoxam resulted in significantly fewer aphid numbers than did imidacloprid-treated plants after 3 weeks, likely due to faster imidacloprid metabolism in the plant, which would result in reduced insecticide activity over a shorter time [27]. However, the difference between imidacloprid and thiamethoxam observed in the laboratory trials was not observed in the field trial, possibly because the final CPB larval field count was conducted on 25 June. Also, the effect of high amounts of simulated rainfall may be more pronounced in the greenhouse pots than under actual field production.

Similarly, the percentage leaf tissue consumed by larvae was significantly (P<0.05) higher with dinotefuran and the untreated than with imidacloprid or thiamethoxam on all dates recorded. However, dinotefuran resulted in significantly less leaf tissue consumption than was observed with plants that received no insecticide in late May and early June (Fig. 2) only under the low rainfall regimen; there were no significant differences between dinotefuran and plants that received no insecticide under the high rainfall regimen, again indicating that rainfall level impacted the effectivess of dinotefuran more so than that of imidacloprid or thiamethoxam.

Data from the mortality and leaf-feeding laboratory bioassays showed significant (P<0.05) insecticide effects in that both imidacloprid and thiamethoxam, applied at planting with the seed piece, were significantly (P<0.05) more effective in reducing foliage consumption by CPB larvae from planting (23 Apr) through 7 Jul under both the low and high rainfall regimens compared with plants that were treated with dinotefuran or plants that received no insecticide. The interaction showed that plants treated with dinotefuran, under the low rainfall regimen, were significantly more effective in reducing leaf-feeding damage than plants that received no insecticide only from planting through 7 Jun; dinotefuran was ineffective in reducing foliage consumption under the high rainfall regimen for all dates observed as compared with plants that received no insecticide. Similarly, leaves from plants treated with either imidacloprid or thiamethoxam at planting resulted in significantly higher CPB mortality from May through mid-July under the low rainfall regimen and through mid-June under the high rainfall regimen as compared with CPB mortality on plants treated with dinotefuran or on plants that received no insecticide. As observed with the leaf-feeding bioassay, CPB mortality was significantly higher on plants treated with dinotefuran under the low rainfall regimen only through early June, as compared with plants that received no insecticide; CPB mortality on plants treated with dinotefuran under the high rainfall regimen were not significantly different than the untreated on all dates observed. A highly water-soluble systemic insecticide may not provide long-term control or regulation compared with a less water-soluble systemic insecticide [28]. Mortality of CPB gradually on foliage treated with imidacloprid or thiamethoxam declined over time from May through July for both rainfall regimens, and mortality declined more quickly with dinotefuran for the low rainfall regimen as compared with imidacloprid or thiamethoxam.

Data from the field trials for CPB mortality are similar to the CPB mortality trials in the laboratory bioassay, with significant insecticide effects in that plots treated with either Admire or Platinum at planting resulting in significantly fewer CPB/3 hills through late June under both the low and high rainfall regimens. Plots treated with dinotefuran had significantly fewer CPB/3 hills than did plots that received no insecticide only on 2 Jun and only under the low rainfall regimen; there were no significant differences for CPB/3 hills between plots treated with dinotefuran and the no-insecticide plots on all dates observed under the high rainfall regimen. As was recorded with the CPB larval counts in the field and the laboratory trials, dinotefuran was less effective than either imidacloprid or thiamethoxam under the low rainfall regimen, and dinotefuran was ineffective under the high rainfall regimen. Further, both imidacloprid and thiamethoxam remained effective for a greater time period after planting than did dinotefuran, possibly due to the lower solubility of imidacloprid and thiamethoxam. Field data for the percentage defoliation caused by CPB feeding were similar to that observed for CPB mortality. Significant insecticide effects were observed as defoliation increased over time. Defoliation in plots treated with either imidacloprid or thiamethoxam remained low through all observation dates, but defoliation in plots treated with dinotefuran was not significantly different than plots that received no insecticide on any observation date. These data agree with [29], who reported that potatoes treated with dinotefuran had significantly greater percentage defoliation than either imidacloprid or thiamethoxam in white potatoes with in-furrow treatments.

The PLH data show significant insecticide and significant rainfall effects. Under the low rainfall regimen, all insecticides resulted in significantly less PLH damage through the season (mid- June through mid-July) as compared with the plots that received no insecticide; no significant effects among the three insecticides were observed for either observation date. However, under the high rainfall regimen, plots treated with either imidacloprid or thiamethoxam resulted in significantly less ($P<0.05$) PLH damage early in the season (25 Jun) than did all other treatments, but only plots treated with thiamethoxam had significantly ($P<0.05$) less PLH damage than all other treatments on 7 July. Our data agrees with that of [30], who reported that thiamethoxam provided longer and more consistent protection of snap beans from PLH than did imidacloprid when applied as a seed treatment. They concluded that thiamethoxam had greater physiological activity against PLH than did imidacloprid.

Overall, the effectiveness of dinotefuran against CPB and PLH was influenced by the amount of rainfall significantly more than either imidacloprid or thiamethoxam. Under a high rainfall regimen, plots that received a treatment of dinotefuran were not significantly different from plots that received no insecticide for either CPB mortality or for CPB leaf consumption, and under a low rainfall regimen the effectiveness of dinotefuran against CPB was significantly shorter over time than either imidacloprid or thiamethoxam. It has to be mentioned that our findings reflect the worst-case scenarios for rainfall (maximum rainfall total of 60.96 cm [24"] laboratory and 50.8 cm [20"] field) and that under practical conditions lower amounts of rainfall would likely naturally occur. These high rainfall amounts likely caused exaggerated leaching of all of the neonicotinoids, especially the highly soluble dinotefuran, resulting in reduced control of the CPB and PLH.

Results of the present study show that rainfall has an impact on effectiveness of soil-applied imidacloprid, dinotefuran and thiamethoxam for both CPB and PLH control in white potatoes. However, growers should not rotate imidacloprid or dinotefuran with thiamethoxam as part of their resistance management program. Alyokhin et al. (31) showed that the correlation for $LC_{50}$ values for imidacloprid and thiamethoxam was highly significant using diet incorporation bioassays, and that there was substantial cross-resistance among three neonicotinoid insecticides. Further, Mota-Sanchez et al. (32) reported that cross-resistance was observed with all 10 different neonicotinoids in a bioassay using topical applications. They concluded that the rotation of imidacloprid with other neonicotinoids may not be an effective long-term resistance management strategy. For effective insecticide resistance management, control of CPB should not depend exclusively on the neonicotinoid class of insecticides. It is important to use all available effective pest management tools, including crop rotation, border treatments, and non-neonicotinoid (different class) insecticides.

## Author details

Gerald M. Ghidiu
*Rutgers- the State University, Department of Entomology, New Brunswick, NJ, USA*

Erin M. Hitchner
*Syngenta Crop Protection, Elmer, NJ, USA*

Melvin R. Henninger
*Rutgers- the State University, Department of Plant Biology and Pathology, New Brunswick, NJ, USA*

## 5. References

[1] Capinera, J.L. 2001a. Handbook of Vegetable Pests. Academic Press, New York, NY. pg. 96-98.

[2] Hare, J.D. 1980. Impact of defoliation by the Colorado potato beetle on potato yields. J. Econ. Entomol. 73: 369-373.

[3] Alyokhin, A., M. Baker, D. Mota-Sanchez, G. Dively and E. Grafius. 2008. Colorado potato beetle resistance to insecticides. J. Pot. Res. 88: 395-413.

[4] Capinera, J.L. 2001b. Handbook of Vegetable Pests. Academic Press, New York, NY. pg. 338-340.

[5] Peterson, A.G. and A.A. Granovsky. 1950. Relation of *Empoasca fabae* to hopperburn and yields of potatoes. J. Econ. Entomol. 43: 484-487.

[6] Pienkowski, T.L. and R.T. Medler. 1964. Synoptic weather conditions associated with long-range movement of the potato leafhopper, *Empoasca fabae,* into Wisconsin. Ann. Entomol. Soc. 57: 588-591.

[7] Taylor, P.S. 1993. Phenology of *Empoasca fabae* (Harris) (Homoptera: Cicadellidae) and development of spring-time migrant source populations. Ph.D. dissertation. Cornell University, Ithaca, NY.

[8] Dively, G.P., P.A. Follett, J.J. Linduska and C.K. Roderick. 1998. Use of imidacloprid-treated row mixtures for Colorado Potato Beetle (Coleoptera: Chrysomelidae) management. J. Econ. Entomol. 91: 376-387.

[9] Prabhaker, N., S.J. Castle, S.E. Naranjo, N.C. Toscano and J.G. Morse. 2011. Compatibility of two systemic neonicotinoids, imidacloprid and thiamethoxam, with various natural enemies of agricultural pests. J. Econ. Entomol. 104: 773-781

[10] Elbert, A., M. Haas, B. Springer, W. Thielert and R. Nauen. 2008. Applied aspects of neonicotinoid uses in crop protection. Pest. Manag. Sci. 64: 1099-1105.

[11] Cassida, J.E. and G.B. Quistad. 1997. Safer and more effective insecticides for the future. *In:* D. Rosen [ed], Modern Agriculture and the Environment, Kluwer Academic, UK., pp. 3-15.

[12] Maienfisch, P., M. Angst, F. Brandl, W. Fischer, D. Hofer, H. Kayser, W. Kobel, A. Rindlisbacher, R. Senn, A. Steinemann and J. Widmer. 2001. Chemistry and biology of thiamethoxam: a second generation neonicotinoid. Pest. Manag. Sci. 57: 906-913. doi: 10.1002/ps365

[13] Sur, R. and A. Stork. 2003. Uptake, translocation and metabolism of imidacloprid in plants. Bulletin of Insectology 56: 35-40.

[14] Gawlik, B.M., N. Sotiriou, E.A. Feicht, S. Schulte-Hostede and A. Kettrup. 1997. Alternatives for the determination of the soil adsorption coefficient Koc, of non-ionic organic compounds. Chemosphere 34:2525-2551.

[15] Walker, A. 2000. A simple centrifugation technique for the extraction of soil solution to permit direct measurement of aqueous phase concentrations of pesticide. *In:* J. Cornejo and P. Jamet [ed.], Pesticide/soil interactions – some current research methods. Institute National De La Recherce Agronomique, Paris, pp. 173-178.

[16] Toscano, N.C., B. Drake, F.J. Byrne, C. Gispert and E. Weber. 2007. Laboratory and field evaluations of neonicotinoid insecticides against the glassy-winged sharpshooter. *In:*

Proceedings of the Pierce's Disease Research Symposium, pp. 98-100. San Diego, CA. December 2007.

[17] Harris, C.R. and B.T. Bowman. 1981. The relationship of insecticide solubility in water to toxicity in soil. J. Econ. Entomol. 74: 210-212.

[18] Cloyd, R.A., K.A. Williams, F.J. Byrne and K.E. Kemp. 2012. Interactions of light intensity, insecticide concentration, and time on the efficacy of systemic insecticides in suppressing populations of the sweetpotato whitefly (Hemiptera: Aleyrodidae) and the citrus mealybug (Hemiptera: Pseudococcidae). J. Econ. Entomol. 105: 505-517.

[19] Gupta, S., V.T. Gajbhinge and R.K. Gupta. 2008. Soil dissipation and leaching behavior of a neonicotinoid insecticide thiamethoxam. Bull. Environ. Contam. and Tox. 80: 431-437.

[20] Miles. 1993. Environmental fate of imidacloprid. Miles Report No. 105008 to Department of Environmental Protection, Sacramento, CA. Miles, Inc. 8 pp.

[21] Wakita, T., N. Yasui, E. Yamada and D. Kishi. 2005. Development of a novel insecticide, dinotefuran. J. Pestic. Sci. 30: 122-123.

[22] Brockman, F.E., W.B. Duke and J.F. Hunt. 1975. A rainfall simulator for pesticide leaching studies. Weed Sci. 23: 533-525.

[23] SAS Institute. 1989. User's guide: statistics. SAS Institute, Cary, NC.

[24] Ferro, D.N., Q. Yuan, A. Slocombe and A.F. Tuttle. 1993. Residual activity of insecticides under field conditions for controlling the Colorado potato beetle (Coleoptera: Chrysomelidae). J. Econ. Entomol. 86: 511-516.

[25] Zehnder, G. 1986. Timing of insecticides for control of Colorado potato beetle (Coleoptera: Chrysomelidae) in eastern Virginia based on differential susceptibility of life stages. J. Econ. Entomol. 79: 851-856.

[26] Lu, W., X. Shi, W. Guo, W. Jiang, Z. Xia, W. Fu and G. Li. 2011. Susceptibilities of *Leptinotarsa decemlineata* (Say) in the north Xinjiang Uygur autonomous region in China to two biopesticides and three conventional insecticides. J. Agric. Urban Entomol. 27: 61-73.

[27] Magalhaes, L.C., T.E. Hunt and B.D. Siegfried. 2009. Efficacy of neonicotinoid seed treatments to reduce soybean aphid populations under field and controlled conditions in Nebraska. J. Econ. Entomol. 102: 187-195.

[28] Cloyd, R.A. 2010. Sucking it up: understanding how systemic insecticides kill insect pests. Western Nursery Landscape Association e-Newsletter, 1 Feb 2010. www.wnla.org/news_list.php?artic_idx=883

[29] Kuhar, T.P., H.B. Doughty, E. Hitchner and M. Cassell. 2007. Control of wireworms and other pests of potatoes with soil insecticides. *In*: Arthropod pest management research on vegetables in Virginia. VPI&SU Eastern Shore AREC Report #306: 19.

[30] Nault, B.A., A.G. Taylor, M. Urwiler, T. Rabasey and W.D. Hutchison. 2003. Neonicotinoid seed treatments for managing potato leafhopper infestations in snap bean. Crop Protection 23: 147-154.

[31] Alyokhin, A., G. Dively, M. Patterson, C. Castaldo, D. Rogers, M. Mahoney and J. Wollam. 2007. Resistance and cross-resistance to imidacloprid and thiamethoxam in the Colorado potato beetle, *Leptinotarsa decimlieata* (Say). Pest Manag. Sci 63: 32-41.

[32] Mota-Sanchez, D., R. Hollingworth, E.J. Grafius and D.D. Moyer. 2006. Resistance and cross-resistance to neonicotinoid insecticides and spinosad in the Colorado potato beetle, *Leptinotarsa decemliata* (Say) (Coleoptera: Chrysomelidae). Pest Manag. Sci. 62: 30-37.

Chapter 10

# Heavy Metal Content in Bitter Leaf (*Vernonia amygdalina*) Grown Along Heavy Traffic Routes in Port Harcourt

Ogbonda G. Echem and L. G. Kabari

Additional information is available at the end of the chapter

http://dx.doi.org/10.5772/55604

## 1. Introduction

The herb known as bitter leaf *(Vernonia amygdalina)* is a shrub or small tree that can reach 23 feet in height when fully grown. Bitter leaf has a grey or brown coloured bark, which has a rough texture and is flaked. The herb is an indigenous African plant; which grows in most parts of sub-Saharan Africa. The East African country of Tanzania is traditionally linked to this plant and can be found growing wild along the edges of agricultural fields. It is a medicinal plant and fresh bitter leaf is of great importance in human diet because of the presence of vitamins and mineral salts (Sobukola et al., 2007).

It is a very important protective food and useful for the maintenance of health and prevention and treatment of various diseases. Some principal chemical constituents found in bitter leaf herb are a class of compounds called steroid glycosides- type vernonioside B1 – these chemical substances possess a potent anti-parasitic, anti-tumour, and bactericidal effect. The bitter leaf is mainly employed as an agent in treating schistsomiasis, which is a disease caused by parasitic worms. It is also useful in the treatment of diarrhoea and general physical malaise.

Remedies made from bitter leaf are used in treating 25 common ailments in sub- Saharan African, these include common problems such as fever, and different kinds of intestine complaints, as well as parasite-induced diseases like malaria. Bitter leaf also helps to cleanse such vital organs of the body like the liver and the kidney. Bitter leaf is also used in the treatment of skin infections such as ringworm, rashes and eczema. However, bitter leaf and other vegetables contain both essential and toxic metals over a wide range of concentrations (Radwan and Salama, 2006).

## 2. Heavy metal

This is a member of an ill-defined subset of an element that exhibit metallic properties, which would mainly include the transition metals, some metalloids, lanthanides and actinides. Many different definitions have been proposed - some based on density, some based on atomic number or atomic weight and some on chemical properties or toxicity (Purseglove, 1977)

The primary anthropogenic sources of heavy metals are point sources such as mines, foundries, smelters, coal-burning power plants, and other sources such as combustion of wastes, and vehicle emissions. Some investigations carried out have shown that heavy metal accumulation in vegetables may pose a direct threat to human health.

Vegetables ingest metals by absorbing them from contaminated soil as well from deposits on different parts of the vegetable exposed to the metal from polluted environments. Many heavy metals act as biological poisons even at parts per billion (ppb) levels.

The distribution of heavy metals (such as iron, copper, manganese, lead, chromium and zinc) in leaves, stems and roots of fluted pumpkin (*Telfeiria occidentalis*) were investigated at three different street roads in Cross River State of Nigeria (Edem et al., 2009). The streets were Afokang, Anantigha and Eneobong. The result showed that, the concentrations of Fe, Mn, Pb and Cr were highest in the leaves.

Also, Afokang Street with the heaviest vehicular traffic out of the three streets recorded the highest concentration of Pb in the leaves of the plants. This may be attributed to the high level of exhaust emissions from vehicles plying the road.

Ademoronti (1986) in his findings on a study carried out in Benin, Nigeria, showed that lead deposits on bark of trees were found to vary according to traffic volumes; areas of high traffic volume recorded higher concentrations of lead. Further studies have revealed that lead does not readily accumulate in the fruiting part of vegetables and crops (e.g., corn, beans, tomatoes, apples and berries); a higher concentration of lead is most likely to be found in leafy vegetables and on the surface of roots crops.

According to Ademoronti (1986), plants accumulate minerals essential for their growth from the environment and also accumulate metals such as cadmium (Cd) and chromium (Cr) which have no known direct benefit to the plants (Brook and Robinson 1989). From Ademoroti's (1995) findings, cadmium is present in low concentrations in vegetable leaves, entering the human body through diet. The mechanism of trace metals' movement within plants was investigated by Walsh (1971). He found that most metals become chelated by a relatively simple agent in xylem sap.

The iron content of normal plant tissue varies with the plant species, but it is usually between the range of 20 - 200mg/kg dry matter (Walsh, 1971). The cobalt content of a normal plant is usually within the range of 0.01 – 1.00mg/kg dry matter (Walsh, 1971). Lead is toxic to crops at concentration range of 3 -20 ppm depending on the plant species; and to animals at a concentration of 1mg/day (Bowen, 1979). Zinc is an essential mineral involved in catalytic functions and is important for both man and plants health and growth (Jeffery, 1992). The zinc content of normal plants tissues varies according to plant species, but it is usually within the range of 5 – 300mg/kg dry matter (Walsh, 1971).

Generally vegetables appear to have the highest and lowest amount of heavy metals accumulated in their leaves and seed respectively e.g. beans, peppers, tomatoes, melons and peas show very low intake of heavy metals in their seed. Plant intake of heavy metals varies with soil $P^H$. A study by Echem (2010) on cassava cultivated on oil polluted soil showed that soil contaminated with heavy metals cause contamination of foodstuffs.

Heavy metals are associated with myriad adverse health effects, including allergic reaction, nephrotoxicity, and cancer. Humans are often exposed to heavy metals in various ways - mainly through the inhalation of metals in the workplace or polluted neighbourhoods, or through the ingestion of food that contains high levels of heavy metals or paint chips that contain lead Ifon, (1977).

The three heavy metals commonly cited as being of the greatest public health concern are cadmium, lead and mercury. There is no biological need for any of these three heavy metals. Exposure to cadmium can result in emphysema, renal failure, cardiovascular disease and perhaps cancer. The primary adverse health effect from exposure to lead is neurological impairment (particularly in children). Other adverse health effects associated with lead include sterility in males and nephrotoxicity.

Heavy metals have been reported to play positive and negative roles in human life (Slavesk et al., 1998; Divrikli et al., 2003; Dundar and Saglam, 2004). Some heavy metals like cadmium, lead and mercury are major contaminants of food supply and may be considered as very harmful to the environment since they do not biodegrade while others like iron, zinc and copper are essential for biochemical reactions in the body (Zaidi et al., 2005). Jarup (2003) and Sathawera et al., (2004) have reported that, most heavy metals are not biodegradable, have long biological half-lives and have the potential for accumulation in the different body organs leading to unwanted side effects. There is a strong link between micronutrient of plants, animals and humans, and the uptake and impact of contaminants in these organisms (Yuzbas, et al., 2003; Yaman et al., 2005).

Research findings by Divrikli et al. (2006) had shown that the concentration of essential elements in plants is conditional; it is affected by the characteristics of the soil and the ability of plants to selectively accumulate some metals. Sources of heavy metals for plants include rainfall in atmospheric polluted areas, heavy traffic as a result of high discharge of exhaust effluents- indiscriminate disposal of oil or fossil fuels by road side mechanics, - plant protection chemicals and fertilizers which could be absorbed through leaf blades (Kovacheva et al., 2000; Lozak et al., 2002; Sobukola et al., 2006).

The aim of this research is to ascertain the level of heavy metals ingested by bitter leaf grown along heavy traffic routes. It is a well established fact that diesel, gasoline, lubricants, vehicle parts such as carburettors among others, contain heavy metals. The routes chosen for this study record the highest traffic density for heavy duty vehicles, lorries, buses and cars in Rivers State; this is due to the presence of Onne Sea Port, Eleme and Port Harcourt Refineries, Eleme Petrochemicals among others on one hand, and the movement of goods and persons between Aba and Port Harcourt on the other. On these routes, it is common to see vehicles under- going various types of repairs, mechanic workshops and heavy exhaust emissions, and these invariably become the primary source of pollution to the road side soils. Symptoms associated with acute oral zinc dose are vascular shock, vomiting, diarrhoea, pancreatitis and damage of hepatic parenchyma. Consequently, crop plants growing on heavy metal contaminated medium can accumulate high concentrations of trace metals causing serious health risks to consumers.

## 3. Hypothesis

$H_O$: There is significant difference in the mean concentration of heavy metals in bitter leaf cultivated along heavy traffic route from the ones cultivated away from the routes.

$H_A$: There is no significant difference in the mean concentration of heavy metals in bitter leaf cultivated along heavy traffic routes from the ones cultivated away from the routes.

### 3.1. Materials and methods

The materials used for this study include among others bitter leaf, stainless steel knife, laboratory grinder, polythene bags, desiccator, analytical balance, platinum crucible, muffle furnace, hydrochloric acid, volumetric flask, distilled water and atomic absorption spectrophotometer.

### 3.2. Study area

The research was carried out in three different heavy traffic vehicular routes in Rivers State. These routes are: Eleme Junction – Aba Road, Eleme Junction – Akpajo Road and Akpajo – Refinery Junction. The control location was Kina Gbara Street in Bori, Khana Local Government Area.

### 3.3. Sample collection and preparation

The bitter leaf samples were obtained from the four stated locations. The bitter leaf were harvested from two different sites per location; and then put into separate polythene bags and labelled accordingly. They were then taken immediately to the laboratory for preparation and analysis.

The bitter leaf (*Vernonia amygdalina)* samples were washed with tap water and thoroughly rinsed with distilled water, then dried in an oven at 105°C. They were then pulverized to fine powder using a laboratory grinder. The fine powder was put into polythene bags and preserved in the desiccator. 3.0g of each sample was accurately weighed into clean platinum crucible and ashed at 450-500°C, then cooled to room temperature in the desiccator. The ash was dissolved in 5ml of 20% hydrochloric acid and the solution was carefully transferred into a 100ml volumetric flask. The solution was filtered using Whatman No 1 filter paper (Umoren and Onianwa, 2005) into a 50ml volumetric flask and made up to the mark with distilled deionised water. The samples were taken to the Shimadzu atomic absorption spectrophotometer (model 6650) for aspiration.

The determination of heavy metal (Fe, Zn, Cr, Pb and Cd) content of the sample solution was carried out in accordance with the procedure of the AOAC (1984) on dry samples.

## 4. Results and discussion

The results of the heavy metal analysis of the bitter leaf samples are shown in table 1.

| Sample | Location | Fe | Pb | Cr | Cd | Zn |
|---|---|---|---|---|---|---|
| Bitter Leaf | PH-Aba Road | 2.96 | 0.220 | 0.030 | ND | 0.890 |
| Bitter leaf | Akpajo-Eleme Road | 3.43 | 0.405 | 0.052 | ND | 1.08 |
| Bitter leaf | Refinary-Akpajo Road | 3.27 | 0.350 | 0.041 | ND | 1.12 |
| Mean | | 3.22 | 0.325 | 0.042 | ND | 1.03 |
| Bitter leaf | Kina Gbara Street | 2.21 | 0.156 | 0.024 | ND | 0.84 |

ND = not detected.

**Table 1.** Concentration (mg/kg) of heavy metal in bitter leaf

The results in table 1 show that the concentrations (mg/kg) of the heavy metals fall within the following range: iron (Fe) is 2.21 to 3.43, lead (Pb) is 0.156 to 0.405, chromium (Cr) is 0.024 to 0.052, zinc (Zn) is 0.840 to 1.12, while Cadmium (Cd) was not detected in any of the locations.

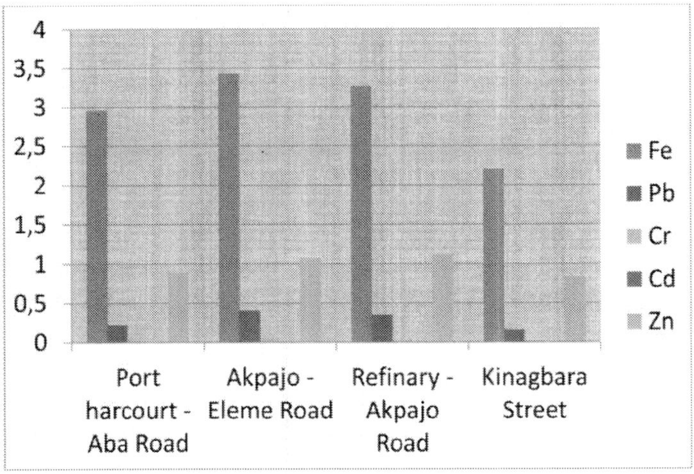


**Figure 1.** Bar Chart for the concentration of heavy metals

The order of concentration of each heavy metal in the bitter leaf harvested from all the studied locations is: iron (Fe) in bitter leaf harvested along Akpajo - Eleme Junction Road soil > Refinery - Akpajo Junction Road soil > Port Harcourt – Aba Road soil > Kina Gbara Street soil (Bori); lead (Pb) in bitter leaf harvested along Akpajo Road soil – Eleme Junction Road soil > Refinery – Akpajo Junction Road soil > Port Harcourt – Aba Road soil > Kina Gbara street soil (Bori); chromium (Cr) in bitter leaf harvested along Akpajo-Eleme Junction Road soil > Refinery – Akpajo Junction Road soil > Port Harcourt Aba Road soil > Kina Gbara street(Bori).

Zinc (Zn) in Refinery – Akpajo Junction Road soil > Akpajo – Eleme Junction Road soil > Port Harcourt – Aba Road soil > Kina Gbara street soil (Bori). The results further reveal that; the high concentration of iron (Fe), lead (Pb), chromium (Cr) and Zinc (Zn) found in the leaves of the bitter leaf may be due to the high levels of exhaust emission from the vehicles plying these roads, (Edem et al., 2009).This view is further supported by the relative lower concentrations of the heavy metals found in the control sample collected in Bori. The level of concentration of Fe in the control sample in addition to the low concentration of toxic heavy metals makes it suitable for human consumption (Ifon, 1977). The high concentration of iron in bitter leaf might be attributed to the nature of the soil and sometimes by the presence of some bacterial that depends on $Fe^{2+}$ for source of energy.

The results of the study indicate that the bitter leaf harvested along Akpajo – Eleme Junction Road had the highest concentration of iron (Fe), lead (Pb) and chromium (Cr), compared to the one from Bori (control), while Refinery-Akpajo Junction Road had the second highest concentration of iron (Fe) lead (Pb) and chromium (Cr). The highest concentration of zinc (Zn) was also recorded along this route.

The findings of this study agree with the results of Edem et al. (2009) on levels of heavy metals in pumpkin leaves harvested in streets with heavy vehicular traffic in Cross River State and the report on the investigation of the deposit of lead on the bark of trees planted in areas of high traffic volume in Benin carried out by Ademoronti (1986).

The mean concentration of Pb (0.325) and Cr (0.041) in the sampled bitter leaf were above the WHO tolerable limit of 0.005 – 0.1 and 0.005 – 0.01 respectively. The bioaccumulation of these heavy metals has adverse health implications to man, especially in children and pregnant women (Dupler, 2001).

The high concentration of Fe reported in this study is in conformity with that published by Hart et al. (2005) on concentrations of trace metals (Pb, Fe, Cu and Zn) in crops harvested in some oil prospecting locations in Rivers State. The likely reason given for this high value of iron is the participation of green vegetables in the synthesis of ferrodoxin. However, the mean concentration of Fe (3.22) falls within the acceptable range (1.0 – 4.0mg/100g) as published by Platt (1980). The mean concentration of Zn (1.03) is also within the acceptable limit. However, from WHO (1984) report, excessive intake of Fe and Zn is capable of causing vomiting, dehydration, electrolytic imbalance and lack of muscular co-ordination.

By using chi-square $x^2$ , i.e.,

The expected value,

Where VT = Vertical Total
HT = Horizontal Total
GT = Ground Total.

| Sample ID | Fe | Pb | Cr | Zn | Total |
|---|---|---|---|---|---|
| B. leaf (Ph Aba Road | 2.96 (2.860) | 0.220 (0.289) | 0.030 (0.036) | 0.890 (0.915) | 4.10 |
| B" Akpajo Etc. | 3.43 (3.465) | 0.405 (0.350) | 0.052 (0.044) | 1.08 (0.582) | 4.967 |
| B" Refinery and Akpajo Road | 3.27 (3.335) | 0.350 (0.337) | 0.041 (0.042) | 1.12 (1.067) | 4.781 |
| **Total** | **9.66** | **0.975** | **0.123** | **3.09** | **13.848** |

, ,

, =0.350,

, ,

, ,

To find the chi-square,

=

.

= 0.0035 + 0.0004 + 0.0013 + 0.0165 + 0.0086 + 0.0005 + 0.001+ 0.0015 + 0.0000 + 0.0007 + 0.4261 + 0.0026

$X^2 = 0.4601$.

At 1% significant level i.e., 0.01

1 – 0.01 = 0.99

d.f = (n-1) (m-1) = (3–1) (4–1)

= (2) (3) = 6

Degree of freedom = d.f = 6

$X^2$ 0.99, 6 = 0.872

Decision: since $X^2 = 0.4601$ is less than the critical value – 0.872, the null hypothesis (HO) is accepted and the alternative hypothesis (HI); is rejected.

Therefore, there is significant difference between B leaf (Bori) control and $X^2 = 0.4601$ at 1% significant level.

## 5. Conclusion

This study has shown that the mean concentration of heavy metals in bitter leaf harvested along heavy traffic route soil is significant compared with the one harvested away from these routes. The observed common practice of cultivating vegetables in the soil along heavy traffic routes in Port Harcourt will in the long run endanger consumers' health since the ingested heavy metals bioaccumulates in the human body. Consequently, there is a need for all the relevant government and non- governmental agencies to bring this to the knowledge of the general public.

## Author details

Ogbonda G. Echem
*Department of Science Laboratory Technology, Rivers State Polytechnic, Bori, Nigeria*

L. G. Kabari
*Department of Science Laboratory Technology, Rivers State Polytechnic, Bori, Nigeria*
*Member, IEEE, Department of Computer Science Rivers State Polytechnic, Bori, Nigeria*

## 6. References

Ademoroti, C.M.A. (1996). Environmental Chemistry and Toxi Ecology, Folulex Press Ltd. Ibadan

Ademoroti, C.M.A. (1986). Environmental Chemistry and Toxi Ecology, Folulex Press Ltd. Ibadan

AOAC (1984). Official methods of analysis of the Association of Official Agricultural Chemists -12 Edn, AOAC, Washington, DC.

Bowen, H.J.M. (1979). Environmental Chemistry of Element Acad Press Inc Ltd, London pp.213-273.

Brooks, R.R. and Robinson, B.H. (1998). Aquatic Phytoremediation by Accumulator Plants CAB International, Wallingford pp.203-226.

Divrikli U., Saracoglu S., Soylak M., Elci L. (2003). Determination of trace heavy metal contents of green vegetable samples from Kayseri-Turkey by flame atomic absorption spectrometry. Fresenius Environ. Bull 12:1123-1125

Dupler, D. (2001). Heavy metals poisoning. Gale Encyclopedia of alternative medicine, Farmington Hills, pp. 23-26

Dundar M.S., and Saglam H.B. (2004). Determination of cadmium and vanadium in tea varieties and their infusions in comparison with infusion processes. Trace Ele. Elect 21:60-63

Edem C.A., Dosunmu I., Miranda I. and Bassey F., (2009). Distribution of heavy metals in leaves, stems and roots of fluted pumpkin (*Telfeira occidentalis)* Pak J. Nutrit, 8:222-224.

Hart A.D., Oboh C.A., Barimalaa I.S. and Sokari T.G. (2005). Concentrations of trace metals (lead, iron, copper and zinc) in crops harvested in some oil prospecting locations in Rivers State, Nigeria. African Journal of Floor and Nutritional Sciences Volume 5 no. 2

Ifon, E.T. (1977). The nutrient composition of some Nigerian leafy green vegetables and physiological availability of their iron content. Ph.D thesis, Department of Biochemistry University of Ibadan, Nigeria.

Jarup L. (2003) Hazards of heavy metals contamination. Br. Med. Bull 68:167-182

Jeffery, P.K. (1992). Environmental toxicology, Edward Arnold Ltd. London. 68-78.

Kovacheva P., Djingova R., Kuleff I. (2000). On the representative sampling of plants for multi-element analysis. Phyto. Bal. 6:91-102

Lozak A., Sotyk K., Ostapezuk P., Fijalek Z. (2002). Determination of selected trace elements in herbs and their infusions. Sci. Total Environ. 289:33-40.

Platt B.S. (1980). Tables of Representative Values of Foods Commonly Used in Tropical Countries. Medical Research Council. Special Report Series. No 302 (Revised Edition of SRS 253) London: Her Majesty's Stationery Office.

Purseglove J.W. (1977). Tropical crops dicotyledons (Volumes I and 2 combine). Longman, London.

Radwan M.A, and Salama A.K. (2006). Market basket survey for some heavy metals in Egyptian fruits and vegetables. Food Chem. Toxicol 44:1273 - 1278.

Sathawera N.G., Parikish D. J., Agrwal Y.K. (2004). Essentials heavy metal in environmental samples from western Indian. Bull Environ Cont. Toxical. 73:756 -761.

Slavesk R., Spirevska I., Stafilov T., Ristor T.(1998). The content of trace metals in some herbal teas and their aqueous extracts. Acta.Pharm. 48:201-2009.

Sobukola O. P., Dairo O. U. (2007). Modeling drying kinetics of fever leaves *(Ocimum viride)* in a convective hot air dryer Niger Food J. 25(1):145-153.

Sobukola O. P., Dairo O.U., Sanni L.O, Odunewu A. V., Fafiolu B.O. (2006). Thin layer drying process of some leafy vegetables under open sun. Food Sci. Technol. Int. 13(1):35-40

Umoren, I. U., and Onianwa, P.C.(2005). Concentration and distribution of some heavy metals in urban soil of Ibadan, Nigeria. Park J. Ind. Res., 48:397 – 401.

Walsh, L.M. (1971). Instrumental method of analysis of soil and plant tissues. Soil Science Society of America, Inc. Madison Wisconsin, USA. 26-30.

WHO (1984). Guideline for Drinking Water Quality. Health Criteria and Supporting Information 2: 63-315.

Yaman M., Okumus, N., Bakirdere, S., Akdeniz I. (2005) . Zinc Speciation in soil and relation with its concentration in fruits, Asian J. Chem. 17:66 -72.

Yuzbas N., Sezgin E., Yildirim M. Yildirim Z, (2003). Survey of lead, cadmium, iron, copper and Zinc in kasar cheese. Food Add. Cont. 20: 464-469.

Zaidi M. I., Asrar A., Mansoor A., Faroogui M.A. (2005) . The heavy metal concentration along roadsides of Quetta and its effects on public health J. Appl. Sci.5(4): 708-711.